Non-Combat Roles
for the
U.S. Military
in the
Post-Cold War Era

Non-Combat Roles for the U.S. Military in the Post-Cold War Era

Edited by

James R. Graham

1993

National Defense University Press
Fort Lesley J. McNair
Washington, DC 20319

National Defense University Press Publications

To increase general knowledge and inform discussion, NDU Press publishes monographs, proceedings of University-sponsored symposia, and books relating to U.S. national security, especially to issues of joint, combined, or coalition warfare, peacekeeping operations, and national strategy. The Press occasionally publishes out-of-print defense classics, historical works, and other especially timely or distinguished writing on national security, and it is the home of *JFQ: Joint Force Qaurterly*, a forum for examining joint and combined warfare and exchanging ideas of importance to all services.

NDU Press publications are sold by the U.S. Government Printing Office. For ordering information, call (202) 783-3238 or write to the Superintendent of Documents, U.S. Government Printing Office, Washington, DC 20402. To order a year's subscription to JFQ, send a check for $22.00 ($27.50 for foreign) or provide your VISA or MasterCard number and expiration date to Superintendent of Documents, PO Box 371954, Pittsburgh, PA 15220-7954. You may FAX your order to (202) 512-2233.

First printing, August 1993

Contents

Foreword

One of the most serious issues being examined within the national defense community is that of increasing the military's non-combat roles and missions. The argument for doing so is that the end of the Cold War has made it possible for the military to assume more non-combat roles and missions. This argument suggests that the realignment of U.S. strategy from a global to a regional orientation calls for increasing such activities as nation building, humanitarian assistance and training, disaster relief, and many aspects of peacekeeping.

The counter argument warns that too great a shift in roles and missions because of this "peace dividend" may leave the military less well prepared to meet the major crises that might develop, for example, with China or the former Soviet Union. This counter argument points to how a strong, combat-ready U.S. military, abetted by responsive international structures like NATO, has deterred aggression since World War II.

The National Defense University Symposium titled "Non-Traditional Roles for the U.S. Military" provided a forum for examining those two arguments and a full spectrum of issues falling between them. A very distinguished group of military and civilian commentators discussed these matters insightfully and extensively. This volume offers the major presentations delivered at the Symposium as well as summaries of ensuing panel discussions. More than just an ordinary proceedings of the Symposium, it can serve as an informative reader for general audiences or a resource book for classes studying this important subject.

PAUL G. CERJAN
Lieutenant General, US Army
President, National Defense
University

Preface

In December 1992, the National Defense University sponsored a symposium entitled *Non-Traditional Roles for the U.S. Military in the Post-Cold War Era*; this symposium examined current thinking on the uses and execution of non-traditional roles and missions. The perception at that time was that as the traditional threat of the previous half-century lessened, and because military forces needed to refocus their efforts, there was an increased emphasis on domestic as well as overseas non-combat functions for the United States armed forces. During the symposium, five panels examined the role of the U.S. military in the following major areas: deterring and containing regional conflict; enhancing stability; supporting democracy; supporting the civil authorities at home and abroad; and capabilities in search of missions.

Symposium attendees discussed a variety of emerging roles including peacekeeping; security assistance; nation building; humanitarian and civil assistance; engineering support; environmental and infrastructural support; border control; counternarcotics; arms control; the education and training of foreign military forces; joint exercises; disaster relief; and population evacuation.

A large number of key government personnel took part in the five panels. Panelists included General Wayne Downing, now commander of the Special Operations Command; Lieutenant General (ret) John Yeosock, Ground Forces Commander in the Persian Gulf War; Colonel Ralph Cossa, Chief Policy Division, United States Pacific Command; Brigadier General (ret) Indar Rikhye, Senior Advisor for United Nations Affairs at the United States Institute of Peace; David Scheffer, currently Special Assistant to the U.S. Ambassador to the United Nations; Colonel Douglas Fraser, Military Advisor to the Canadian Mission to the United Nations; Colonel Robert Seitz, U.S. Army Plans and Policy Division; Enid C.B. Schoettle, Senior Vice President, United Nations Association; Toby Trister Gati, now a Special Assistant to the President for National Security Affairs; Colonel Anthony Ramos, Director, Strategic Policy and Plans, United States Southern Command; Lieutenant General Teddy Allen, Director of the Defense Security Assistance Agency; Brigadier General David Mark, Director of the Nigerian National War College; Mr. Reginald Brown, Assistant Administrator, Near East Bureau, Agency for International

Development; Dr. Anthony Gray, Jr., OSD Director of Global Affairs; Lieutenant General (ret) H.J. Hatch, former head of the U.S. Army Corps of Engineers; Rear Admiral Richard Appelbaum, U.S. Coast Guard Chief of Law Enforcement and Defense Operations; Major General Daniel Hernandez, Commander, California National Guard 40th Division; Major General John Heldstab, U.S. Army Director of Military Support; Admiral Paul David Miller, Commander-in-Chief, United States Atlantic Command; Dr. Alberto Coll, Principal Deputy Assistant Secretary of Defense for Special Operations and Low-Intensity Conflict; Lieutenant General Norman Ehlert, U.S. Marine Corps Deputy Chief of Staff for Plans, Policies and Operations; Rear Admiral Robert Kramek, Chief of Staff, United States Coast Guard; Lieutenant General John B. Conaway, Chief of the National Guard Bureau; and General George Joulwan, Commander-in-Chief, United States Southern Command. Although the panelists did not write papers, a rapporteur was assigned to each panel to capture the information discussed.

This volume includes the collection of rapporteur reports reflecting the presentations, conclusions and recommendations made by the panelists. Retired Army Colonel Harry Summers, currently a syndicated columnist, Ambassador Samuel Lewis, now Assistant Secretary of State for Policy Planning, Ms. Elizabeth Ondaatje, RAND Corporation, Mr. Seth Cropsey, Director of the Asian Studies Program at the Heritage Foundation, and Mr. John Collins, Library of Congress served as moderators. Mr. Tim Sisk, a Fellow at the United States Institute of Peace, Mr. John Brinkerhoff, a consultant on national security matters, and Colonel David Graham and Lieutenant Colonel Marne Peterson, students at the National War College, also helped to prepare the rapporteur reports for three of the panels.

In addition to the panel reports, edited transcripts of Dr. Samuel Huntington's keynote lecture and General John Galvin's banquet talk are included as the first and last chapters of the book. Huntington's lecture sets the stage for the chapters to follow, and Galvin's work offers some sobering thoughts for future consideration on the subject of non-traditional roles for the military.

One of the major findings from this symposium was that we had mistitled the conference. Nearly all panelists and participants argued that the word *"Non-traditional"* was wrong. Indeed the U.S. military had been performing these non-traditional roles ever since their origination over 200 years ago. In reality, since the beginning of World

War II and continuing until the end of the cold war, combat roles and missions have been the focus of the U.S. military. In light of many examples to prove this position, the consensus of the participants at this symposium was that we really discussed *"non-combat"* roles. These are now returning as the primary mission focus of our armed forces today. A related additional finding from this symposium was that the United States Armed Forces should not lose sight of their primary reason for being there—to protect the national interests of the United States. Repeatedly during the conference, panelists and participants alike emphasized the importance of remembering that all of the non-combat missions under discussion were ultimately in support of protecting our national interests. Even events which on the surface appear to be purely humanitarian in nature, such as disaster relief, are indeed training opportunities for the defense mission.

With these thoughts in mind, the ideas, presentations, debates, and discussion items collected here should serve as a foundation for future thinking. They represent a large cross section of thinking at the time of the symposium, and in fact continue to be valid today.

Non-Combat Roles
for the
U.S. Military
in the
Post-Cold War Era

Keynote: Non-Traditional Roles for the U.S. Military

Samuel P. Huntington

I AM DELIGHTED TO HAVE THIS OPPORTUNITY to meet with you this morning at the beginning of this conference on non-traditional roles for the U.S. military. My pleasure is enhanced by the thought, as General Cerjan suggested, that the subject of this conference is a non-subject.

Few issues are more important than the missions and functions of the U.S. military in the post-Cold War world. We are in the midst of a major change in the structure of the international system and in the challenges confronting U.S. foreign and national security policy. This is, however, not the first time we have confronted such challenges. At the beginning of the republic, we had to create military and naval forces to deal with threats from European powers. With the end of the Napoleonic Wars, our national security problems changed dramatically and so also did our military policies and forces. These remained fixed in their essentials until the end of the century when we emerged on the world scene as a major power.

Samuel P. Huntington is Eaton Professor of the Science of Government and Director of the John M. Olin Institute for Strategic Studies at the Center for International Affairs, Harvard University. Dr. Huntington is the author or editor of over a dozen books and ninety scholarly articles. He has studied, taught, and written widely in three principal areas: military politics, strategy, and civil-military relations; American and comparative politics; and, political development and the politics of less developed countries.

As a result we consigned the Indian-fighting army and the commerce-protecting navy to history and instead created an army designed for big wars and a navy designed for big battles. This system served us well in two world wars. And then in the late 1940s we had to create a new military establishment for the Cold War. Now that conflict is over, and we are once again confronting the need to debate the nature of our national interests and the missions of our armed forces, just as we did in 1784, 1815, 1898, and 1946. In effect we have to move into the fifth phase of American military policy.

The subject of this conference is the "non-traditional roles" of the armed forces in this phase, which, obviously, implies that there is a distinction between traditional and non-traditional roles. The traditional roles will presumably continue, but in this fifth phase of American military history, the armed forces will also perform new, non-traditional roles. Many of these possible roles are listed in the program of this conference. Others have been promoted by members of Congress, particularly by Senator Sam Nunn. Largely as a result of his leadership, the 1993 Defense Authorization Act encourages the military to conduct a drug demand reduction outreach program targeted at inner-city youths, to provide "role models for United States youth," to furnish "health care to residents of medically underserved communities," and to cooperate with civilians "in addressing domestic needs" and working "to improve the environment and economic and social conditions." In his speech on the Senate floor, Senator Nunn said:

> While the Soviet threat is gone, at home we are still battling drugs, poverty, urban decay, lack of self-esteem, unemployment, and racism. The military certainly cannot solve these problems.... But I am convinced that there is a proper and important role the armed forces can play in addressing these pressing issues. I believe we can reinvigorate the military's spectrum of capabilities to address such needs as deteriorating infrastructure, the lack of role models for

tens of thousands, if not millions, of young people, limited training and education opportunities for the disadvantaged, and serious health and nutrition problems facing many of our citizens, particularly our children.[1]

These would clearly seem to be non-traditional roles. But are they really? The more I have thought about this issue, the more I have become convinced that with one possible exception the subject of this conference does not exist. There are almost no conceivable roles for the American military in this new phase of national security that the American military have not performed in some earlier phase. The true distinction, I believe, is not between traditional and non-traditional roles but between military and non-military roles, or more precisely, between combat missions of the military and the non-combat uses of military force. The purpose of military forces is combat, that is to deter and to defeat the enemies of the United States; that is their central mission, their *raison d'etre*, the only justification for expending resources on their creation and maintenance. The forces created for that mission, however, can and throughout our history have been employed in non-combat non-military uses.

I will not go deep into this history, which I am sure, is familiar to you, but let me just highlight a few of the traditional, non-combat uses of the armed forces in our history. For over three decades West Point trained all of this country's engineers, civilian as well as military. Throughout the nineteenth century, the Army engaged in economic and political development. It explored and surveyed the West, chose the sites for forts and settlements, planned settlements, built roads and developed waterways. For years, the Army performed the functions later assumed by the Weather Bureau and the Geological Survey. In the latter part of the century, the Signal Corps pioneered in the development and use of the telegraph and telephone. The Navy was equally active in exploration and scientific research. U.S. naval ships explored the Amazon, surveyed Latin American and North American coasts, laid cables, collected scientific specimens and data from around the world. They policed the slave trade. Naval officers negotiated dozens of treaties and served in the lighthouse service, the life-saving service, the coastal survey, and the steamboat inspection service. The Army ran civil government in the South during Reconstruction and simultaneously for ten years also governed Alaska. It was, of course, frequently called upon to deal with strikes and labor violence. The Corps of Engineers constructed public buildings and canals and other public works including the Panama Canal. Soldiers helped deal with malaria in Panama and cholera, hunger, illiteracy in Cuba, Haiti, and Nicaragua. They also established schools, built public works, promoted public health, organized elections, and encouraged democracy in these countries. In the 1930s, the Army took on the immense task of recruiting, organizing, and administering the Civilian Conservation Corps.

This past year after the hurricanes in Florida and Hawaii many people hailed the superb contribution the military made to disaster relief as evidence of a "new role" for the U.S. armed forces. Nothing could be more off the target. The U.S. military have regularly provided such relief in the past. As the official U.S. military history puts it, during the 1920s and 1930s, "the most conspicuous employment of the Army within the United States... was in a variety of tasks that only the Army had the resources and organization to tackle quickly. In floods and blizzards and hurricanes it was the Army that was first on the spot with cots, blankets, and food."[2] That has been the case throughout our history. It is hard to conceive of any non-military role for the U.S. military that does not have some precedent in U.S. history. Non-military functions of the armed forces are as American as apple pie.

Throughout our history also, however, these non-military uses of the armed forces have never served as the justification for the maintenance of armed services. The overall size, composition, organization, recruitment, equipment, and training of the armed forces have been justified by the needs of national security and the military missions, the combat missions, which the armed forces may have to perform. In this new fifth phase of American international relations, the security and military functions of the armed forces remain as important as ever. They are the reasons why we will and should continue to have military forces in the coming years.

How can these military missions be defined? There are, I think, at least three such missions.

First, for the first time in sixty years, no major power, no peer rival, poses a security challenge to the United States. It is obviously in our interests to maintain this situation, a goal which has been affirmed by both Presidents Reagan and Bush and in the initial version of the Defense Guidance issued in the winter of 1992. We now need the military policy and forces not to contain and deter an existing threat as we did during the Cold War but rather to prevent the emergence of a new threat. To accomplish this goal, we must maintain substantial invulnerable nuclear retaliatory forces plus military deployments in Europe and Asia for reassurance and to preclude rearmament by Germany or Japan. We must also maintain U.S. technological superiority and U.S. maritime superiority, and provide the base for the rapid and effective creation of a new enhanced military capability if a major power threat should begin to emerge.

Second, regional powers do pose significant threats to American interests in Southwest Asia and East Asia, and we must have the capability to deal with those threats as we did in the Gulf War. To deter or defeat regional aggression the United States will need light and heavy ground forces, tactical aviation, naval and marine forces designed to fight from the sea against land targets, and the sea and air lift to deploy ground forces rapidly to the scene of combat. Ideally the U.S. should be able to fight the equivalent of another Gulf War. The Bush Administration's Base Force and Secretary of Defense Les Aspin's Option C both purportedly would provide this capability. Whether five years from now American forces will in fact be able to fight another Gulf War against an Iraq-like enemy remains to be seen.

Our decisive victory in the Gulf War, however, makes it unlikely that we will be able to repeat that victory in the future. Any major regional aggressor in the future is likely to have and to use nuclear weapons. This point is reflected in the response of the Indian defense minister when asked what lesson he drew from the Gulf War. He replied: "Don't fight the United States unless you have nuclear weapons."[3] The most likely aggressors of the future—North Korea, Iran, Iraq, and others—are intent on acquiring nuclear weapons. Until they do have these weapons, the probability of stability in their regions is reasonably high. Once they do have these weapons, the likelihood they will use them will be high. In all likelihood the first sure knowledge the world will have that they possess a usable nuclear weapon will be the explosion of that weapon on the territory of a neighbor. That act would be coupled with a massive conventional offensive designed to produce the quick occupation of Seoul, the Saudi oil fields, or whatever other target the aggressor had in mind. This is the most serious type of regional threat the U.S. may have to confront, and it may well also be one of the most probable.

Coping with such aggression will place new demands, and what in the context of this conference we might call non-traditional, demands on U.S. military forces. They will have to fight an enemy who has a small number of nuclear weapons and little or no inhibitions about using them. To deter this first use by a rogue state, the United States will have to threaten massive, including possible nuclear, retaliation against such a state. The central function of the Strategic Command in the coming years will be maintenance of nuclear peace in the Third World.

Third, the U.S. military may also have to intervene quickly and effectively in countries important to the U.S. in order to support a friendly government, restore a friendly government that has been overthrown, overthrow a hostile regime, protect American lives and property, rescue hostages, eliminate terrorists, destroy drug mafias, and engage in other actions which normally fall under the heading of "low intensity conflict". Whether or not a state is aggressive or pacific, reasonably decent or totally threatening, depends overwhelmingly on the nature of its government. President Clinton has very appropriately said that the promotion of democracy should be a central, perhaps the central, theme of American foreign policy. In areas critical to its security, the U.S. has to be prepared to defend governments that are friendly and democratic and to overthrow governments that are unfriendly and undemocratic.

This requirement also emphasizes a new role for American military forces: the targeting of dictatorial governments and their leaders. In the Gulf War the U.S. military degraded by more than fifty percent the capability of the Iraqi forces. The U.S. military also substantially brought Iraqi society to a standstill. The U.S. military, however, proved incapable of eliminating the true villain of the piece: the Iraqi government. The elimination of Saddam Hussein was a U.S. objective, although not one endorsed by the United Nations, and we failed to achieve that objective. During the past decade, indeed, we have tried to eliminate three hostile dictators: Khadaffi, Noriega, and Saddam Hussein. We only succeeded with Noriega, and it took us some while and we suffered some embarrassment in doing that, although it involved a minuscule country with respect to which our intelligence had to be better than almost any other place in the world. Targeting and incapacitating dictatorial governments will be an important mission for the U.S. military in the coming years, and it is one with respect to which our capabilities are now sadly deficient.

These are some of the military missions of the United States armed forces in the post-Cold War world. What then are appropriate non-military or civilian uses to which the armed forces may be put in this world? As I have indicated, historically the non-combat activities have been multiple and diverse. There is no reason why they should not be so in the future.

First, these uses of the armed forces could well include those domestic functions in American society highlighted by Senator Nunn and included in the Defense Authorization Act.

Second, these uses obviously should include humanitarian relief at home and abroad where such activity is welcomed by the local government.

Third, peacekeeping at the invitation of the parties involved is clearly also an appropriate activity for the U.S. armed forces.

Fourth, there is the activity about which there may be more questions and which has been highlighted by the crises and the tragedies in Bosnia and Somalia. Should the U.S. armed forces be used to provide humanitarian relief in situations where such efforts are likely to be opposed by one or more of the conflicting parties? Clearly some form of international authorization, presumably from the UN, is a prerequisite to U.S. action. This occurred in the precedent-breaking UN Security Council Resolution 688 that authorized intervention by American, British, and French forces to protect the Kurds in Iraq. The UN has also approved the deployment of outside military forces to assist with the provision of humanitarian relief in Bosnia and Somalia.

The goal of our involvement in such situations presumably is to insure that relief supplies reach their intended beneficiaries. This means that American military forces should be able to take any military action necessary to prevent or to eliminate hostile action against the relief efforts. That is fine. The need, however, is to define the limits of American involvement, and this gives rise to two problems.

First, so long as the country concerned remains in a state of civil war or anarchy, external military forces will be required to ensure that the food and medical supplies are delivered to their intended recipients. If the United Nations is unable to provide those forces, this could mean an extended if not indefinite commitment of American troops. This would not be a Gulf War type situation where we could drive the invading forces out of Kuwait and then pack up and go home. It could mean waiting for the South Slavs or other conflicting parties to resolve their political problems before we could extricate ourselves. And that could be a very, very long wait.

Second, there is the problem of becoming involved in the politics of the country. One or more parties to the conflict sees outside involvement as a hostile act. By coming in, we then, from their viewpoint, become the enemy. Inevitably while we are there for humanitarian purposes our presence has military and political consequences. The United States has a clear humanitarian interest in preventing genocide and starvation, and the American people will within limits support intervention to deal with such tragedies. If Somali clans and Slavic groups are fighting each other, we will attempt to mitigate whatever horrendous consequences may flow from such violence. In such circumstances the American people may even accept some American casualties. The United States has no interest, however, in which clan dominates Somalia or where the boundary lines are drawn in the former Yugoslavia. The American people will not support military intervention which appears to be directed towards such political goals. It is morally unjustifiable and politically indefensible that Americans should be killed in order to prevent Serbs and Bosnians from killing each other.

The U.S. armed forces can and should, as appropriate, be put to a variety of civilian uses, including social renewal and economic development at home, humanitarian and disaster relief at home and abroad, and peacekeeping abroad. The American armed forces should perform military missions involving possible conflict, however, only when they promote American security and are directed against the enemies of the United States.

The possible non-military functions of the armed forces have recently received a great deal of attention. Arguments have been made that the American military should be organized and trained so as to perform such functions. The proposal has been floated, for instance, that a unified command for humanitarian operations should be created. In somewhat similar fashion, a commission of former government officials has proposed establishing a new military command headed by a three or four-star officer to provide support for UN operations and to develop doctrine, to carry out planning, and to train U.S. forces for UN missions. The United States, another group has argued, "should retain and promote officers whose expertise includes peace-keeping, humanitarian administration and civilian support operations..." [4]

Such proposals are, I believe, basically misconceived. The mission of the American armed forces is to combat, to deter, and to defeat the enemies of the United States. The American military should be recruited, organized, trained, and equipped solely for that task. Military forces should, where appropriate, be used in humanitarian and other civilian activities, but they should not be organized or prepared or trained to perform such functions. The core purpose of a military force is fundamentally anti-humanitarian: it is to kill people in the most efficient way possible. It is only for that and related purposes that this country and other countries maintain military forces. Should the military perform other functions? Absolutely yes, and as I have indicated they have performed such functions throughout our history. Should these other functions define the mission of the military? Absolutely not. They should be spill-over functions which the military is capable of providing because they have been well organized, trained, and equipped to perform their military functions of defending this country against its enemies.

The criterion for judging individual military programs and people should be their contribution to this military mission. As we move into Phase Five of American defense policy, concentration on military purpose and military effectiveness is more important than ever. The military budget has been decreasing and will continue to decrease, probably at an accelerating rate. This is in some measure desirable and in large measure unavoidable. The danger is not that defense budgets will go down but that those portions of the defense budget that are devoted to necessary military purposes will go down faster than the budget decline as a whole while programs and activities that have political support and are socially popular will be retained. The military should willingly undertake whatever non-military functions the President and Congress may assign to it.

If the military is to maintain itself in the coming years, however, it must define its purposes purely in military terms. That is its only defense against those who would prostitute it to serve other ends and to promote their own social and political goals. The American tradition in the past has been to maintain military forces solely for military purposes but to make those forces available as needed for constructive civilian uses, and that should be the American practice in the future.

Notes

1. P.L. 102-484, 28 October 1992, Sections. 376, 1045, 1081; Senator Sam Nunn, *Congressional Record*, Vol 138 (no. 91, 23 June 1992), p. S 8602.
2. Maurice Matloff, General Editor, *American Military History* (Washington: United States Army, Office of the Chief of Military History, 1969), p. 413.
3. Quoted in Les Aspin, "*From Deterrence to Denuking: Dealing with Proliferation in the 1960s*," Memorandum, 18 February 1992, p. 6.
4. Carnegie Endowment for International Peace and Institute for International Economics, *Memorandum to the President-Elect, Subject Harnessing Process to Purpose* (Washington: Carnegie Endowment for International Peace, 1992), p. 17; Thomas G. Weiss and Kurt M. Campbell, "*Military Humanitarianism,*" *Survival*, Vol. 33 (September-October 1991), p. 457.

Evolving Roles for the U.S. Armed Forces in Regional Deterrence and Containment

Harry G. Summers, Jr.
Moderator and Rapporteur

$\mathbf{W}$HAT ARE THE EVOLVING U.S. MILITARY roles in regional deterrence and containment in the post-Cold War era? This question was the focus of the symposium's first session, setting the stage for more detailed discussions later in the conference. The issue was examined from three perspectives.

First, Air Force Colonel Ralph A. Cossa, Chief, Policy Division, U.S. Pacific Command, addressed the value of "cooperative engagement," the adaptive forward presence of U.S. military forces in a theater of operations, as a major factor in peacetime regional deterrence and wartime crisis response.

Next, Army Lieutenant General Wayne A. Downing, Commander of the U.S. Army Special Operations Command discussed the unique role played by special operations forces,

---

Harry G. Summers, Jr. is a syndicated columnist for the Los Angeles Times, editor of Vietnam magazine, and a consultant and lecturer on foreign and military affairs. He was a military analyst for NBC News during the Persian Gulf War and has made numerous television appearances, including ABC, CBS and NBC News. Mr. Summers' award-winning critique of the Vietnam War, On Strategy, is used as a student text by the war and staff colleges and by many civilian universities.

especially Army special forces, civil affairs and psychological operations units, in both peacetime deterrence and coalition warfare.

Finally, retired Army Lieutenant General John J. Yeosock, commander of the Third U.S. Army during the Persian Gulf War, which included command of the British and French military contingents, addressed the changing nature of coalition warfare in the post-Cold War era and its importance for future U.S. military operations.

Cooperative Engagement

It is most appropriate that the U.S. Pacific Command (PACOM) be represented at this symposium as we discuss the evolving roles for the U.S. Armed Forces in regional deterrence and containment. The Pacific Command has been the epitome of an "economy of force" theater even before budget cuts made it necessary elsewhere. And when it comes to evolving post-Cold War roles, the U.S. strategy for the Asian-Pacific area has already evolved.

In 1990, under Congressional mandate, the Department of Defense formulated *A Strategic Framework for the Asian-Pacific Rim: Looking Toward the 21st Century.* Twice updated since that time, the 1992 report reflects a consensus between the Congress and the Administration on the fundamental precepts of Asian-Pacific strategy.

Congress found that "The alliances between the United States and its allies in East Asia greatly contribute to the security of the Asian-Pacific region; It is in the national interest of the United States to maintain a forward-deployed military presence in East Asia; . . . and, Finally, and most importantly, the United States remains committed to the security of its friends and allies around the Asia-Pacific rim."

These documents provide the foundation upon which the Commander-in-Chief, Pacific Command (CINCPAC), formulates and implements his military strategy in the Pacific. Called "cooperative engagement," it is founded on a process of adaptive forward presence.

"Forward presence" is a cumulative term. It includes *forward stationing*, the actual stationing of U.S. military forces on the ground at U.S. military installations, as in Japan and Korea today. *Forward deployment* includes these forces, plus embarked Marine Air-Ground Task Forces, Navy carrier battle groups, and Air Force squadrons on temporary deployment abroad. Finally, *Forward Presence* includes all of the above plus ship visits, security assistance exercises, and the wide variety of ways the U.S. military interacts with the Asian community.

The objectives of "cooperative engagement" are straight-forward; engagement and participation during peacetime, deterrence and cooperation in crisis, and unilateral or multilateral victory in conflict.

Cooperative engagement is built on three pillars: preserving existing alliances and friendships; maintaining a forward presence in order to demonstrate America's continuing commitment to regional security and stability; and ensuring the ability, if necessary, to react to crises.

Currently, five security alliances—with Australia, Japan, Korea, the Philippines and Thailand—provide the framework both for deterrence and for military interaction with America's Asian-Pacific neighbors and underscore as well the continuing U.S. commitment to remain a Pacific and a global power. Complementing these alliances is a broader concept of a cooperative international community among nations that share common security interests with the United States. The U.S. will continue to develop both bilateral and multilateral ties in order to establish this partnership of nations. Because of the tremendous diversity in Asia, much of this effort will be accomplished, at least initially, on a bilateral basis.

In Northeast Asia, U.S. defense policy centers around the security alliances with Japan and Korea. Maintaining these two bilateral alliances remains key to any cooperative engagement strategy. U.S. forces situated in both countries represent the bulk of America's forward-stationed presence in Asia-Pacific.

The U.S.-Japan security relationship has been called the single most important bilateral relationship in Asia, if not in the world. It not only directly affects the security interests of the world's two largest economic superpowers, but directly affects the national security interests of virtually every other country in the Asia-Pacific region.

To the west, North Korea's future intentions and direction, especially with regard to nuclear weapons, provide cause for concern. North Korea remains one of the most closed, repressive, heavily-armed, and economically backward societies in the world. The U.S. presence in South Korea has been instrumental in preserving peace on the peninsula. It has provided the security shield behind which the Republic of Korea's military forces have developed and Korea's economy has blossomed. The gradual improvement in the Republic of Korea's military capability has set the stage for a measured reduction in the level of U.S. presence. However there are no plans either to withdraw completely from Korea or to diminish in any way the importance of the U.S.-Korea alliance.

One of the great unknowns in Asia is China's future direction once the current gerontocracy passes from the leadership scene. China is indeed a nation in the midst of a profound transition. While future U.S. policy toward China awaits articulation, it is clear that China must be engaged, not isolated, and encouraged to interact on a responsible basis with all its Asian neighbors, the U.S. included.

To the north, the U.S. is expanding its military-to-military contacts with Mongolia. This military relationship focuses on educational assistance, medical training, and building a logistics capability and infrastructure so that the nation's vast natural resources can better serve the needs of the people.

Finally, on the periphery of northeast Asia, the Russian Republic still controls one of the largest standing military forces and a strategic nuclear capability and arsenal second to none. Given the size of its military and a national history of expansionism that predates the Bolshevik Revolution by several centuries, to say that the Russians no longer represent a *potential* threat is to deny both history and reality.   Nonetheless, the opportunity to engage the Russians constructively has never been better. On a military-to-military basis, U.S. efforts to interact with Russia in the Far East will concentrate on confidence-building measures and on increased cooperation in settling or defusing potential crises.

U.S. defense policy in Southeast Asia centers around the maintenance of existing security alliances with the Philippines and Thailand and on the continuing improvement of existing friendships with other ASEAN states. Despite the withdrawal of U.S. forward-stationed forces from the Philippines, the treaty still provides for combined training and exercises, periodic ship visits and routine aircraft transits.

The U.S. enjoys routine access to Thai military installations and facilities and has a vigorous military exercise program which promotes confidence and interoperability.   The other ASEAN states, to varying degrees, all appear supportive of a continued U.S. presence and the stability it provides.

At the high end of the spectrum is the agreement by Singapore to house a modest 135-person logistics support unit to facilitate interaction. More typical are agreements, such as the one with Indonesia, that simply permit U.S. naval ships to use local maintenance and repair facilities. U.S. defense interaction with Vietnam, Laos and Cambodia currently centers on attempts to achieve a full accounting of America's missing-in-action from the war in Vietnam, and the U.S. fully endorses ASEAN and UN attempts to bring peace to Cambodia.

Australia remains a fundamental element in the U.S. security strategy in the Pacific.   The security arrangement remains on solid ground, based on a mutual desire to cooperate on a wide variety of defense issues. The U.S. also has direct defense ties with several South Pacific nations, such as the Freely Associated States of the Republic of the Marshall Islands and the

Federated States of Micronesia as well as the UN Trust Territory of Palua.

U.S. defense policy in the Indian ocean centers around improved military-to-military relations with India. With the world's fourth largest army and seventh largest navy, India is now and will remain the major power in South Asia. Dialogue between the U.S. and Indian militaries has never been better.

The second pillar of the strategy of cooperative engagement is forward presence, the cornerstone of future U.S. Asia-Pacific security strategy, which on a daily basis demonstrates the continued U.S. commitment to remain an Asian power.

In the past, forward presence has been largely synonymous with forward-based forces and the network of bases/facilities that made this possible. This continues to be true in northeast Asia. But forward presence has always included more than just American troops stationed abroad. It also includes naval patrols, ship visits, joint and combined exercises, visits of military personnel, on-site training programs, and personnel exchanges. Another element is the humanitarian assistance and nation-building efforts carried out by the various U.S. military services.

A cooperative engagement strategy employing adaptive presence will not require additional overseas bases. But it will require continued and expanded access to regional ports, airfields, and training areas which support crisis contingency operations, making everything from humanitarian disaster relief to the evacuation of non-combatants during emergencies easier.

This ability to respond—the third pillar in the cooperative engagement strategy—is central to CINPAC's mission of promoting a stable and secure world and essential to the maintenance of ongoing alliances and friendships. Ironically, crisis response capabilities can be accomplished without common agreement on the nature of future threats. Given past track records, it is probably safe to make two predictions: first, more crises will occur; and second, they will not be the specific problems we predicted or prepared for.

Special Operations Forces

The peacetime and wartime roles of special operations forces in the post-Cold War world characterize these forces as "ambassadors in uniform." With the collapse of the Soviet Union, the threat to U.S. national interests has changed. In the past the peace/conflict/war Spectrum of Conflict emphasized *combat operations.* The focus was on direct action, special reconnaissance, unconventional warfare, foreign internal defense, counterterrorism, psychological operations, civil affairs, and coalition warfare.

Now the greatest danger arises from the realities of the underdeveloped "Third World." Socioeconomic decline and environmental instability has led to ethnic strife, the spread of religious fundamentalism, weapons proliferation, resource shortages, and political unrest. In Latin America narco- trafficking has led to narcoterrorism as an instrument of policy.

As a result, while combat operations remain important, the future Spectrum of Conflict will emphasize *peacetime engagement.* The focus will be on maintaining order, separating clashing groups, preventative peacekeeping, active peacekeeping, supervising cease fires, precluding mass migrations and humanitarian assistance.

The U.S. Army Special Operations Command (USASOC) is uniquely qualified for these missions. Some, 30,000 strong, with half of its forces on active duty and half in the reserves, its primary peacetime engagement units include five active and four reserve Special Forces Groups, one active Civil Affairs battalion and three reserve Civil Affairs commands, one active and three reserve Psychological operations groups.

Their normal tempo of operations is intense. On any given day, for example, over 1,000 Army Special Operations forces are deployed overseas in 30-35 different countries on 75-95 different missions. During fiscal year 1992, their overseas deployments in the Atlantic Command (12%) , Central Command (6%) , European Command (20%), Pacific Command (17%) and

Southern Command (45%) included counterdrug activities, training with foreign forces, security assistance and other operational missions in support of the regional commanders-in-chief and of U.S. ambassadors abroad.

Most were Army Special Forces personnel. When deployed, they live and work with the local people and develop a high degree of rapport. Like their Civil Affairs and Psychological Operations counterparts, inter-cultural skills—foreign language ability, cultural and area orientation, and interpersonal non-verbal communication—are what distinguishes them from other line soldiers. All are hand-picked and highly skilled professionals. Officers and non-commissioned officers alike have at least five to eight years service.

Each Special Forces group has a specific regional orientation. Peacetime engagement includes training, advising, and assisting U.S. and allied forces or agencies. In both peace and war U.S. Special Forces develop, organize, equip, train and advise or direct indigenous military and paramilitary forces. As a strategic strike force, combat missions include operations in remote and denied areas for extended periods of time with little external direction or support. In addition, during the Persian Gulf War, Special Forces provided liaison teams to 106 Arab battalions and Navy SEALS assisted in maritime search and seizure operations. Given America's widespread network of military alliances, support to coalitions is an important function. That includes coordinating military and paramilitary operations across the operational spectrum and encouraging military cooperation at all levels of operations. Other tasks include assessing, adapting and integrating coalition tactical doctrines to met operational requirements, and training, advising, assisting and organizing coalition forces. Coalition support varies widely depending on the needs of a coalition, its composition, and the operational environment. Such support is not limited to Special Operations personnel, and other agencies, both military and civilian, may be involved. In any case, however, specialized training is a necessity, for different levels of operations may require specific operational emphasis.

With the growing importance of military forces to humanitarian relief operations, Civil Affairs, yet another capability of Special Operations Forces,, has become increasingly important. Like Special Forces, these Civil Affairs units have a regional orientation. Ninety-five percent of these units are in the Army Reserve, reflecting the special skills needed to accomplish the civil affairs mission.

For many of these personnel—judges, law enforcement personnel, civil engineers, firemen and the like—there is a one-to-one correlation between their civilian skills and their military duties. These tasks include reestablishing civilian control, including public safety, public health and otherwise providing for the public welfare.

While these Civil Affairs units can be mobilized in wartime, for peacetime engagement we must now rely on volunteers. What is needed in legislative authority to call reserve units in small numbers to deal with such eventualities.

The same is true for Psychological Operations (PSYOPS) personnel, eighty-five percent of whom are also in the Army Reserve. There is, however, an active duty PSYOPS battalion, and a multi-media state-of-the-art PSYOPS facility at the U.S. Army Special Operations Command's Fort Bragg, North Carolina headquarters.

PSYOPS units too have a regional orientation, a focus essential for sociological research, target audience analysis, and campaign planning. Such propaganda analysis, development, production, and dissemination are critical to ensure getting the message across.

While we talk today about the "evolving" role for the U.S. Armed Forces in regional deterrence and containment, Special Operations Forces are already involved in such missions. Operations and training deployments during the past 18 months, not including some 15 to 20 deployments that remain secret, involved some 91 different countries ranging from Antigua to

Zimbabwe, including such former Iron Curtain counties as
Bulgaria, the Commonwealth of Independent States [the former
Soviet Union], China, Czechoslovakia, Hungary, Mongolia,
Poland, Romania and Vietnam.

There is no question that the Special Operations Force
community has the capabilities in-being to play a role in deterrence and containment. The pitfalls lie elsewhere.

Most serious is the need for a coherent articulation of
U.S. strategy by the Department of Defense and Department of
State. The regional commanders-in-chief and country teams
around the world, as well as other government agencies, must
know precisely what we are attempting to do. Otherwise we can
never achieve the interagency coordination essential to effectively
bringing the power of the United States government to bear.

Coalition Warfare

While we speak of "non-traditional" roles for the U.S.
military in the post-Cold War era, one "traditional" aspect will
endure. We are clearly going to fight any future conflict in a
coalition context.

One obvious reason is the physical assets—ships, aircraft,
armor, artillery and fighting personnel—that allies contribute to
the struggle. Less obvious, but even more important, is their
psychological impact. In a July 1991 article, The *New Republic's*
Charles Krauthammer pointed out that what we call
"multilateralism" has actually been *pseudo*-multilateralism, with
the United States carrying the majority of the load.

That was true in the so-called United Nations Command
in the Korean War, where at its peak strength UN ground forces
consisted of 590,911 from the Republic of Korea, 302,483 from
the United States, and 39,145 from other UN countries. The
same was true in the Vietnam War, where the peak 1969 strength
of the 68,889 "Free World Military Forces" was dwarfed by the
some 800,000 Republic of Vietnam Armed Forces and the half
million U.S. forces. And while the over 200,000 allied coalition

forces in the Persian Gulf War made a much more significant contribution, they were still overshadowed by the more than 530,000 U.S. forces deployed.

But to some degree those figures are immaterial. As Krauthammer emphasized, "Americans insist on the multilateral pretense. A large segment of American opinion doubts the legitimacy of unilateral American action but accepts action taken under the rubric of the 'world community.'"

But while future conflicts will be fought in a coalition context, such coalitions will differ dramatically from those of the past. Unlike the NATO alliance, for example, which dominated military thinking for the past 45 years, future coalitions will not be standardized and codified, with all differences resolved by memos of understanding.

They will not have the equivalent of a Supreme Allied Commander, Europe or a Commander-in-Chief, United Nations Command in Korea, who exercises unity of command. And the task at hand will not be reinforcing forward deployed forces from the continental United States for a World War III scenario. Yet by and large we are still tied to those strategies of the past, and that is still how we are organized, equipped, manned and trained. Worse yet, too often that is how we still think.

Future coalitions will be ad hoc, put together rapidly, as with the Persian Gulf War coalition, to meet unexpected and unplanned for contingencies. Instead of a single commander coordinating allied efforts we will likely have individual allied force commanders. For unity of command we will have to substitute unity of effort, as we did successfully in the Persian Gulf War.

And the Persian Gulf War highlighted another critical aspect of coalition warfare—the importance of clear-cut agreed-upon political objectives. "No one starts a war," said Carl von Clausewitz in *On War*, "or rather no one in his senses ought to do so, without being clear in his mind what he intends to achieve by that war and how he intends to conduct it.

Failure to heed that admonition was one of the fatal flaws of the Vietnam War. "Almost 70 percent of the Army generals who managed the war," reported Brigadier General Douglas Kinnard, "were uncertain of its objectives." But in the Persian Gulf War President George Bush ensured that the objectives were well understood by the American people, by the military commanders in the field, and by our allies abroad. And it showed.

The U.S. had learned from its Vietnam experience that Clausewitz was right when he emphasized that the value of a war is determined by its objectives, and that value determines the sacrifices to be made in pursuit of it both in magnitude and duration. In World War II the value was national survival, and America sacrificed a million casualties in pursuit of it. In Vietnam the value was never made clear, the cost soon became exorbitant, and public support evaporated..

What is not generally understood it that same logic applies to coalitions as well. The value of the objective is the force that causes alliances to coalesce in the first place, and the cement that holds them together until the crisis is resolved.

In World War II, it was the threat of Adolf Hitler that created the unholy alliance between the Soviet Union and the Western democracies. To defeat the Nazis, said Soviet dictator Joseph Stalin, "I would deal with the Devil and his grandmother," a sentiment shared by British Prime Minister Winston Churchill and U.S. President Franklin D. Roosevelt.

Likewise, it was the threat posed by Iraq's Saddam Hussein which caused the Persian Gulf War coalition of such unlikely Arab bed fellows as erstwhile mortal enemies Saudi Arabia, Syria and Egypt to come together. This threat held them together throughout the course of the war.

When it comes to coalitions, we need to remember the assumptions upon which past coalitions have been built. NATO, for example, did not just happen. It was the threat of Soviet aggression that brought the North Atlantic Treaty Organization together in 1949 and that held it together for the next four decades. It remains to be seen whether, with the dissolution of the Soviet Union, that cement will still hold.

Be that as it may, it is the threat of a common enemy and the need for a common response to aggression that will cause future coalitions to form. The United States Government and its military forces must have the mental and physical flexibility to answer such unanticipated demands, understanding that in most cases, the U.S. will bear the majority of the burden, for it is the only nation that has a force projection capability.

Although the current emphasis on the use of military force to respond to humanitarian concerns, as in Somalia, may be an example of future non-traditional roles, the old verities still apply. What is the U.S. objective in Somalia? Is its value sufficiently strong to ensure the support of the American people, who must bear the cost of such ventures? Is it sufficiently strong to cause other UN members to share and ultimately take over the burden of such relief? And what effect will such deployments have on the war-fighting capabilities of the Armed Forces, which is the primary reason for their existence? These are the kinds of questions that must be addressed as we commit our military to such non-traditional missions. If we do intend to project forces in a timely, meaningful manner to handle those cases where deterrence failed, we need to reexamine our intellectual assumptions. When do we commit U.S. forces abroad? The recent Los Angeles riots may provide an insight. Maintaining internal security, along with protecting the American homeland and safeguarding U.S. interests abroad, is one of the three missions assigned to the Armed Forces by law. In responding to that domestic security mission, however, we have always used a staged response beginning at the lowest levels of force.

The first line of defense is the local police. If they cannot handle it, then the state police are called in. The next step is local National Guard units under State control. Only in extremis, and at the request of State authorities, are Federal troops committed to restore order. That same phased escalatory technique may well apply to commitment of U.S. forces to international security missions as well.

Conclusion

Be it cooperative engagement, special operations forces, or coalition operations, the United States has the military capability to use its Armed Forces in regional deterrence and containment.

One of the reasons is that these forces have been adequately funded over the years. But they are an extremely perishable commodity. Whether the Armed Forces will continue to be funded at adequate levels so that the U.S. retains the capability for such regional deterrence and containment remains to be seen.

While Special Operations Forces have been reoriented to peacetime engagement, the regular Armed Forces have not yet fully made that transition. Warfighting must remain their primary emphasis, but as in Somalia today the men and women of the Army, Navy, Air Force, Marine Corps, and Coast Guard must also be ready to operate globally in non-traditional roles.

Enhancing Stability:
Peacemaking and Peacekeeping

Samuel W. Lewis, Moderator
Tim Sisk, Rapporteur

IN LATE NOVEMBER 1992, THEN-PRESIDENT George Bush ordered the United States military, together with much smaller contingents from other countries, to embark on an unprecedented mission—to militarily intervene in a sovereign state, Somalia, in order to alleviate widespread famine and starvation precipitated by a brutal, two-year civil war. Backed by a United Nations Security Council resolution authorizing the mission, the president ordered the military to secure the delivery of humanitarian relief to millions of starving Somalis and halt the widespread fighting and looting that had prevented these supplies from reaching their destination. Some 25,000 U.S. troops have taken part in this operation, which has a goal of securing the relief routes before handing the torch to the UN in a bid to reconstruct the shattered Somali polity.

The Somalia operation, dubbed Operation Restore Hope,

Samuel W. Lewis is President, United States Institute of Peace. He is a retired Career Minister in the U.S. Foreign Service. After leaving the U.S. Government, Mr. Lewis initially joined the Johns Hopkins University's Foreign Policy Institute in Washington, D.C. as Diplomat-in-Residence. He was also a Guest Scholar at the Brookings Institute in Washington, D.C.

Tim Sisk is a Research Fellow at the United States Institute of Peace.

was unprecedented in two ways. It was the first time the United Nations authorized military intervention in what is essentially an internal conflict to secure the provision of humanitarian relief. And it was the first operation of its kind in which the U.S. military intervened in an internal dispute under UN auspices for these purposes. The Somalia operation thus represents an important shift in the relationship between the United Nations and its member states, and between the United States (and its military) and the world body. These new relationships, however, have been developing for some time, linked in part to the overall changes in global relationships following the end of the Cold War and the new prestige and role thrust upon the UN in this dawning era. The 1991 war in the Persian Gulf, fought to rebuff Iraq's invasion of Kuwait in 1990, was perhaps the first indication of these new relationships. The new international norms reflected by the Gulf War—which have been taken a step further in Somalia—have profoundly altered the means by which the UN seeks to make peace in this era, and the ways in which the United States and its military power are increasingly relied upon to bolster UN initiatives.

Shortly after the end of the UN-backed multilateral effort to oust Iraq in 1991, the United States Institute of Peace established a study group to explore the lessons that might be learned from this experience for the UN system . The group sought to consider how the UN might prepare itself to better deter the kinds of events that led to the need for Operation Desert Storm and to help keep the peace once peace is reestablished. One question that arose very early in the study group's discussions was *Under what terms can an international body composed of more than 160 countries with very disparate ideas about their priorities realistically agree when an international force should intervene in a sovereign state to restore or preserve order?* The possibility of intervention confronts a central issue that the UN, from its beginning, has avoided because built into the UN Charter is a countervailing principle—that of noninterference in the domestic affairs of member states.

This principle is often perceived as a sacred one, particularly for smaller countries in the third world who see in potential UN interventions, even under an international aegis, the seeds of a new era of western colonialism. When the United States takes a leadership role or proposes a course of action involving UN intervention, which is necessary if the UN Security Council is to act successfully, the specter of neo-colonialism immediately arises. This diplomatic issue is one that must be dealt with carefully, even as the military and logistical requirements are planned for an intervention to restore order and permit the delivery of humanitarian assistance to Somalia. Thus, there exists a tension between the principle of humanitarian intervention and state sovereignty.

In the post-Cold War era, the inherent tension between humanitarian intervention and state sovereignty is likely to continue to dominate debate on the appropriateness of UN action in essentially internal conflicts. The increasing role of the United States military in UN Peacekeeping missions will also be argued, especially in light of the fact that following the "passing of the torch" to the UN in the Somalia operation, U.S. troops—perhaps as many as five thousand—will stay deployed as part of the UN peacekeeping mission providing logistical support. So for the first time, the United States will be a substantial contributor to a UN peacekeeping mission. These issues raise a number of questions for the U.S. military, which have been addressed by the panel on "Enhancing Stability: Peacemaking and Peacekeeping" organized and led by the United States Institute of Peace. This chapter offers edited versions of the presentations made by this panel. They have been updated by panel participants to reflect recent events, particularly with regard to Somalia. The remarks reflect the individuals' personal points of view, and not necessarily those of the Institute or the institutions with which the participants are affiliated.

First David Sheffer, a senior associate at the Carnegie

Endowment for International Peace, discusses humanitarian intervention versus state sovereignty. In his presentation, he covers international law as it pertains to the Somalia situation. Next, Major General Indar Jit Rikhye (Ret), Senior Advisor for United Nations Affairs at the United States Institute for Peace, suggests several considerations for creating institutional reform within the United Nations and its peacekeeping operations. Colonel Richard Seitz, Chief of the Strategic Plans and Policy Division at the Department of Army, then shares his views on the role of the United States military and UN peacekeeping. Next, Enid C.B. Schoettle, Director of the Project on International Organizations and Law and Senior Fellow at the Council on Foreign Relations, discusses United Nations financial requirements for peacekeeping operations. And finally, Toby Trister Gati, Special Assistant to the President for National Security Affairs and Senior Director for Russia, Ukraine, and Eurasian Affairs on the National Security Council, closes the discussion by providing some ideas on the role and impact of the United States on UN capabilities.

Humanitarian Intervention versus State Sovereignty

I intend to briefly cover a large waterfront area of international law and principle that bears upon the Somalia operation. This issue of humanitarian intervention is probably the most challenging foreign policy issue confronting President Clinton. Its complexities and its integration of military and political factors are becoming overwhelming. It is the issue of the hour. I will examine humanitarian intervention, which has legal authority that in fact is expanding, and the notion of sovereignty, a legal norm that is contracting. I will finish with a short discussion about the national interest and how we define it in the post-Cold War world in relation to humanitarian intervention and sovereignty.

This issue is now moving so fast that it's very difficult for anyone to sit back and provide a definitive analysis. Over the last year I have had the opportunity many times to do so and events keep outstripping the analysis. Nonetheless, I want to posit some fairly provocative propositions at the outset and then provide more detail.

First, I propose that we are witnessing the end of sovereignty as it has been traditionally understood in international law and in state practice. In its place, we are seeing a new form of national integrity emerging. The term "sovereignty"—because of the baggage it carries over hundreds of years of state practice and evolving international law—is becoming somewhat impractical for describing what needs to be done with respect to certain situations in many states. We need to focus our thinking more on the concept of what protects national integrity, fully recognizing that so many of the activities that take place within state borders, whether humanitarian, environmental, social, or otherwise, have broad international implications that the old notion of sovereignty simply does not address.

We are also witnessing the end of humanitarian intervention as it has been defined and implemented in the past. It still has many of its traditional characteristics, but humanitarian intervention is beginning to shift toward new criteria, which we need to recognize. Humanitarian intervention is now characterized by its multilateral character, the use of armed force if necessary, and even the promotion of political change. Thus, the nonforcible character of humanitarian relief operations is expanding into forcible options. Humanitarian intervention also is becoming a justifiable option in civil wars, which previously were outside the ambit of foreign interference.

Finally, the "national interest" has become a somewhat misleading term, given the criteria and requirements of the post-Cold War world. We should begin to focus on what is in this country's global interests, as well as our national interests, which typically focus only on the requirement for national defense of our territory.

The Notion of Sovereignty

International law has never fully recognized exclusive sovereignty. The whole premise of international law is that states are subject to certain obligations that impinge upon their freedom of action. And yet, from the Treaty of Westphalia in 1648 onward, there has been a fairly clear understanding that sovereignty is a pillar of international law and that everything revolves around it. Sovereignty qualified everything we did in legal practice and particularly in political practice. But I do not want to start in 1648. I want to start in 1945 because one of the main arguments in analyzing the right of humanitarian intervention is Article 2(7) of the UN Charter. This provision appears as an underlying principle of the UN Charter and thus applies to all subsequent clauses:

> Nothing contained in the present Charter shall authorize the United Nations to intervene in matters which are essentially within the domestic jurisdiction of any state or shall require the Members to submit such matters to settlement under the present Charter; but this principle shall not prejudice the application of enforcement measures under Chapter VII.

As you all know, Chapter VII was employed in the 1991 Iraq conflict, and it has been employed with respect to UN operations in both Bosnia and Somalia.

Thus, since 1945 there has been a clear exception for enforcement activities to the principle of noninterference in internal affairs or "domestic jurisdiction." But I would suggest that there are other "carve outs" with which we have considerable experience. There has been much evolution in UN practice and UN law with respect to exceptions to Article 2(7). At Dumbarton Oaks and in San Francisco in 1944 and 1945, diplomats actually sat down and tried to define "interference in the domestic affairs of nations" and what should be tolerated. At the time, the United States argued for the greatest restraint on interference in internal affairs, in contrast to the views of other countries. This position should be kept in mind as we now articulate new forms of intervention for humanitarian purposes. The tables have turned, in some sense, in that today many developing countries are relying upon Article 2(7) to preserve their sovereignty, whereas the United States and some other major powers are beginning to penetrate that veil of sovereignty by justifying intervention on humanitarian grounds.

At Dumbarton Oaks, the principle of noninterference was deemed to apply solely to domestic matters as defined by international law. Who would determine what was a domestic matter? It would be the International Court of Justice. In fact, the U.S. State Department advocated that the International Court of Justice would be the arbiter of such decisions. The White House objected and the State Department backed down, as did Congress. Ultimately, at San Francisco the United States deleted any reference to international law as a basis for determining what is within the ambit of domestic jurisdiction. Furthermore, the United States deleted the word "solely" from the text of Article 2(7) and replaced it with the word "essentially," meaning that the matter can be something other than just solely within a nation's domestic affairs. If a matter is essentially within a nation's domestic affairs, there should be no interference.

Finally, U.S. representatives argued that there is only one entity that determines what constitutes domestic affairs, and that entity is the country involved, particularly if it is the United States. That concern was the major one of U.S. delegates at the time. It arose from the traditional attitude of the U.S. Senate, which had rejected the League of Nations Covenant and U.S. participation in the Permanent Court of International Justice during the 1920s and 1930s. The Senate was concerned that no international body should have anything to do with the domestic affairs of the United States. Thus, the principle of noninterference was injected into the UN Charter in San Francisco, and the United States took the lead in doing so.

The U.S. policy of 1945 need not prevail today. But I think it is very important to understand that this principle has its origins with the U.S. government in 1945, when the language was amended so that it had a fairly broad application to activity within a state's territory. Nonetheless, there is ample authority in the Senate hearings on the UN Charter, suggesting that Article 2(7) was to be interpreted broadly. Secretary of State John Foster Dulles testified as follows: "[Article 2(7)] is an evolving concept. We don't know 15, 20 years from now what in fact is going to be within the domestic jurisdiction of nations. International law is evolving, state practice is evolving. There's no way we can definitively define in 1945 what is within domestic jurisdiction. Let's just let this thing drift for a few years and see how it comes out."

I think Dulles' premonition has been borne out. There is a tremendous amount of UN practice now that has fine-tuned what Article 2(7) means. The UN's approach to South Africa and apartheid had a lot to do with forcing nations to define what is and is not within their respective domestic jurisdictions. For example, it was determined that apartheid has international implications and is not within South Africa's domestic jurisdiction. From practice, one can create a checklist of what, in fact, has been carved out of Article 2(7).

First, the obvious exception is a Chapter VII enforcement action as stated in Article 2(7) itself. There is no dispute in this regard. But one must recognize that over the last 45 years there has been a continued expansion of what is meant by a Chapter VII threat to international peace and security; this expansion includes anything from apartheid to a racist government in Southern Rhodesia to humanitarian catastrophes and atrocities that we now see in play in various parts of the world. Those situations are now falling under the ambit of threats to international peace and security. So the evolution of what constitutes a threat to international peace and security is not one of fact or law, but of definition.

A second exception to Article 2(7) is international treaty commitments. These commitments include, for example, the many international and regional human rights treaties, including the Genocide Convention.

A third exception is anarchy or the absence of consensual authority. If there exists an anarchic situation such as has occurred in Somalia, one cannot credibly argue that there is a government to deal with or that there is consent to be sought. The whole notion of domestic jurisdiction is somewhat poorly articulated in the case of a country like Somalia, where what constitutes domestic jurisdiction is very much a question of fact depending on the vicissitudes of the situation from day to day.

A fourth exception to the principle of noninterference exists when consent would be required from an illegitimate government. A contemporary example is the case of Haiti. Is it really true that the Haitian government, which regional organizations and the UN have now defined illegitimate, can determine what is within the domestic jurisdiction of Haiti and thus prevent intervention in that country's affairs? So a fourth exclusion from Article 2(7) would be situations in which it has been determined by the international community that an illegitimate government exists. In such a case, it may not be necessary to obtain the illegitimate government's consent in order to take the kind of humanitarian action that may be required.

Finally, the fifth and probably most important exception would be a case of systematic violation of the human rights of large groups of people within borders. This condition has been repeatedly articulated as an exception to Article 2(7).

So there are five different exceptions to the principle of noninterference, one expressed in the language of Article 2(7) itself and four through state and UN practice. Today there is, in a sense, a very fluid situation in a number of conflicts in which the limits of sovereignty are increasingly being determined on a case-by-case basis.

Humanitarian Intervention

How is humanitarian intervention itself changing in the context of sovereignty? I address this question by focusing on some of the current conflicts in which these issues feature prominently, namely Bosnia and Somalia.

Bosnia was declared and admitted into the United Nations as the nation-state of Bosnia-Herzegovina following the dissolution of Yugoslavia. Bosnia made appeals for assistance against Serbian aggression, and if anyone looked at the traditional principles of collective self-defense, there would be ample authority under international law to have assisted the government of Bosnia for the purpose of preserving its sovereignty. It was a sovereignty issue as well as a humanitarian one. The specific request by the government of Bosnia went unheeded, and now we are faced in Bosnia primarily with a humanitarian challenge: how do we stop the "ethnic cleansing" and other human atrocities that are occurring there. Ultimately, we have to return to the issue of sovereignty in Bosnia and whether or not a humanitarian intervention will be employed to restore the sovereignty of a member state of the United Nations. If so, how will that sovereignty be overseen by the United Nations and protected?

The situation in Bosnia shows that we are not through with the notion of sovereignty yet, but that the notion is certainly changing in a very dramatic way. Bosnia will present a very, very difficult challenge regarding whether we completely abandon the notion of sovereignty or whether, ironically, we try to preserve it through the instrument of humanitarian intervention.

With respect to Somalia, a vacuum of sovereignty exists. Yet ultimately we know that there is going to have to be a state-building exercise in Somalia following humanitarian intervention.

Thus, the issue arises of whether there will be an intermediate step such a UN trusteeship or conservatorship—as argued in an excellent article by Gerald Helman and Steve Ratner in a recent issue of *Foreign Policy*—or some other kind of interregnum before Somalia regains sovereignty. What sort of sovereignty will it be and under what conditions?

The changing characteristics of humanitarian intervention, upon which I have only touched, are moving toward multilateralism as opposed to unilateralism. There have been many examples of unilateral intervention in the last forty years. Today, the United Nations is moving toward the use of military force to intervene in civil wars, which used to be anathema in international law. By the twenty-first century the principle of nonintervention in civil wars probably will be abandoned. Such intervention in civil wars will be the norm, for humanitarian reasons as well as for state-building and government restoration exercises that one are already being reflected in legal and political developments in the UN, the Organization of American States (OAS), and the Conference on Security and Cooperation in Europe (CSCE), where concepts of democratic government are being promoted.

National Interest

Finally, one brief remark about the national interest. Humanitarian intervention probably will drive the United States to take another hard look at what it conceives to be in the national interest. This look will require some radical rethinking. Humanitarian concerns are not necessarily threats to the borders of the United States, nor are they necessarily threats to most of our major allies around the world. Humanitarian calamities are not necessarily "regional conflicts," which is the term of art that the Pentagon tends to invoke. Rather, humanitarian problems are internal conflicts in which the interest is the survival of large civilian populations, the preservation of whole national economies that, if destroyed, will have a costly impact for decades to come. Such destruction ultimately may affect the United States, the ecosystem (which can be severely endangered when humanitarian calamities occur), and the larger goal of international peace and security. How we define the national interest thus will become increasingly important as we begin to engage more deeply in humanitarian interventions.

UN Institutional Reform and Effective Peacekeeping Operations

These remarks are made within the context of the first summit meeting of members of the Security Council at the heads-of-state and heads-of-government level on January 30, 1992, as well as the Secretary General's response, articulated in his July 1992 report entitled *An Agenda of Peace*. This report will not repeat in detail many of the recommendations, such as the call for strengthening peacekeeping as well as other matters related to managing of conflicts, that have already come forth as a result of the summit meeting. But these are important recommendations that provide a framework for the following discussion of institutional reform.

The Institutional Framework

There are four main UN institutions involved in maintaining international peace and security. The first, as we all know, is the Security Council, which has the primary responsibility for maintenance of international peace and security. The second is the General Assembly, which deals with all matters of peaceful negotiations but may refer specifically to this somewhat reserved area when the Security Council is not seized with that particular issue or when the Security Council refers the matter to the General Assembly, which it has in practice been doing under the Uniting for Peace Resolution of 1954. One of the earliest examples of this is the establishment of the first UN peacekeeping force ever, the United Nations Emergency Force I (UNEF I), after the Suez Crisis. This occurred because of the French and the British vetoes in the Security Council.

The third institution is the International Court of Justice. The fourth is the Office of the Secretary General, which has taken on additional responsibilities as a result of the Cold War. Under the UN Charter, the Secretary General is described as the chief administrative officer who, under Article 99, may bring to the attention of the Security Council matters which threaten international peace and security.

Cooperation among the permanent members of the Security Council has reduced, if not removed, the need for turning to the General Assembly in matters of international peacekeeping. Therefore, there is far less use of the General Assembly in this particular field. However, the General Assembly may discuss these matters when so requested by the Security Council.

Many members—including the United States—do not recognize the binding jurisdiction of the International Court of Justice. Lately, however, it has been put to use, and in the Nicaraguan border case for example, its ruling has been accepted by the parties concerned and presumably by the United States, which has not rejected it. The Secretary General, in *An Agenda of Peace,* has recommended special use of the International Court of Justice for advisory opinions.

The Security Council and the Secretary General

The main operating institutions within the UN system for peacekeeping are the Security Council and the Office of the Secretary General. Their specific relationship is therefore especially important with respect to what is happening in Somalia, the former Yugoslavia, and elsewhere.

With respect to the Security Council, the five permanent members—each of which holds a veto over Council decisions—currently show a great deal of cooperation. It is important, when analyzing the Council's response to a conflict, to study the interrelationship of the five. Such a study would show that China really has considerable hesitation on the question of intervention and agrees only in very specific cases.

For instance, in the case of Somalia, China in fact accepted intervention, although this acceptance has not been fully recognized by the media. Similarly, you will find the nonpermanent members have a view and a majority of them oppose intervention, although they have accepted it under the special exception definitions. There has been a switch from the purity of nonintervention to an acceptance under special, case-by-case circumstances. In the case of the former Yugoslavia, the nonpermanent members and China insisted on consent of the successor states because they had proper governments, even though the old federation had broken into several states. However, in the case of Somalia, the nonpermanent members accepted the fact that there was no authority in Somalia; there was no government, and therefore the UN could intervene if that seemed to be the only way of providing any humanitarian assistance to the people who are suffering. Thus, during the Security Council's meeting on the Secretary General's report recommending intervention, the nonpermanent members who represent the so-called third world were largely supportive of a resolution that permits enforcement action in Somalia specifically for the purpose of providing humanitarian assistance.

The UN Military Infrastructure

When making such decisions about the prospect of a peace enforcement action, the five permanent members have their own large missions to the UN and their own military advisors. Nonpermanent members are sometimes very small states, with small diplomatic staffs that usually are helped out by an ally or a friendly country. Generally, the nonpermanent members on the Security Council have little support and depend largely on either friends or the UN Secretary General to provide necessary support and guidance.

Within the UN Secretariat, the Security Council staff is quite small. It is insufficient to provide the right kind of support to the nonpermanent members who lack sufficient staff resources to manage the number of conflicts with which the UN Security Council deals. The Military Staff Committee is not functioning. (It had three meetings in relation to the sanctions against Iraq, and they would have ended without agreement. Luckily for the committee, its deliberations were overtaken by Desert Storm, so there was no need to continue the meetings.)

The Secretary General has a small executive staff that consists of a chief of staff (a reduced post from what used to be called the *chef de cabinet*, which many thought had become very powerful), and a senior political advisor. The reduced staff can barely serve the Secretary General and is too small to fully coordinate at the highest level all the various activities that go on in the United Nations.

Secretary General Boutros Boutros-Ghali, in his reorganization, has created two under secretary general positions for conflict prevention and resolution and divided the world between them; together they have a staff of about 200. In discussing the reorganization of the Secretariat, both of the under secretaries have said that "the desk staff was recruited during the Cold War and it was meant to do nothing." Thus, the reorganization of the Secretariat involves not only a reduction of numbers but also an improvement in quality.

The Office of the Under Secretary for Peacekeeping has many years of experience in managing peacekeeping operations. There is a long tradition, a long record of service, and a good institutional memory there. But they are very short on staff. The present Under Secretary General manages a total of more than

50,000 troops. There are not many countries that have armed forces of such a size. The present staff of ten is expected to eventually increase to fourteen or fifteen persons.

The administrative and logistics staff, known as Field Service, belongs to another branch, the Office of Administration, Management, and Human Resources. Linkage between peace-keeping and administration offices has been established and they have been able to borrow two or three military logistic officers from Field Service. The civilians—who have basically come up the rank structure—have not received specialized training and lack experience to manage the military logistics staff. The military advisor and a small staff run peacekeeping operations on the same basis as before, when the UN had fewer peacekeeping missions. The commanders of the UN minions receive broad directives; there's a great deal of personal relationship between the Under Secretary General and the commanders; and they have a marvelous way with the people in the field. The unfortunate part is that they simply do not have the resources to manage such large operations.

Another group which has just been established is the Office of the Under Secretary of Humanitarian Affairs. We must allow it a few more months before we are able to comment on it.

What is required in the reorganization that Secretary General Boutros-Ghali, among others, has called for? He has reorganized the major under secretary positions and pledged to reduce the assistant secretary general positions; he is going to do that. He's also pledged to increase the number of women at the senior level to 25 percent. The Secretary General has now in turn called on his under secretaries to make reductions, instead of additions, in certain sectors. This job will be a tough one, as some very serious personnel problems will arise.

Conclusion: The New Peacekeeping Challenges

In conclusion, we will examine the question of the new peacekeeping challenges as exemplified by Somalia and the former Yugoslavia. What are the main problems in peacekeeping and how can we solve them? These problems are already being experienced in the former Yugoslavia. But they are even more evident in Somalia where the UN had been attempting to deploy a peacekeeping force of little more than 3,500. It had succeeded in bringing only a battalion of 500 Pakistani troops, which had not been permitted to go beyond the airport in Mogadishu after several months of negotiations. In other words, the UN troops have not really been permitted to carry out their operational functions.

UN peacekeeping operations are still very hastily put together; they are still "band-aid" affairs. It is not the fault of the Secretariat; rather, it is a packaging problem. There is very little preparation before these operations are deployed. There is the anomalous example of ten years' preparation for the UN operation in Namibia (UNTAG); when suddenly the budget was slashed by half, which necessitated new preparation about six weeks prior to the operation. Not surprisingly, it faced a critical situation when fighting among the feuding parties broke out anew on D-day when they finally opened the mission.

There is a lack of administration and logistics. The commander and the staff meet for the first time at the start of the mission; there has been no previous training; they receive only a very broad mission from the United Nations; and the commander in the field is more or less told, "Well, you do what you have to do." In times of crisis, commanders are usually told to do what they think is best.

Another difficulty is the parties' refusal to cooperate with the UN. They make commitments and pledges, often only to break them. For example, there have been many cease-fires in the former Yugoslavia. When they break down, force is often used against the UN.

The Congo is an example of an operation we should not forget, because there is very little in the current crises that was not present in that mission, with the exception of the change in attitude toward sovereignty and the question of being able to plan ahead. What the Secretary General requires is a task force, similar to the old Congo Club that Dag Hammarskjöld had. What the Secretary General requires is an advisory committee, which he now does not have, to provide the political input by troop-contributing countries and those who are directly interested and involved and have an influence on the conflict. There are committees of troop contributing countries now, but they only talk about logistics and changes of personnel.

A second requirement is that the UN special representative in the field must meet with diplomatic representatives on a regular basis. It appears to be taking place in most instances, but how often is uncertain. There must be enough political officers in the field with the UN peacekeeping operations because they are political operations—a continuation of diplomacy by use of troops without force.

Finally, the Secretary General should have a military advisory staff. Former secretaries general had one, but the staff has now been placed back under the Under Secretary General of Peacekeeping. It was not a very wise step to make this change because it confines military advisors to one particular section. Their advice should also be available to everyone during the planning phase and they should have the ears of the Secretary General.

The U.S. Military and UN Peacekeeping

There have been a lot of dramatic changes among countries in the last few years. The world is free of ideological conflict for the most part, and the prospect of a global war has diminished greatly. But there is a potential for regional conflict. Regional can be defined both as intra- and inter- state, and that potential has grown and clearly become more visible to everyone. In response, the United Nations has become more involved in a wide array of peace operations. The number of operations has grown not only quantitatively, but also qualitatively in the last few years.

During the Cold War, superpower competition kept the United States with most of its military power faced off against the Soviet Union in Europe; any conflicts on the periphery, when they occurred, were either fueled by superpower rivalry or they were contained because we wanted to suppress the conflict from spreading. Peacekeeping operations were limited to conflict containment using third-party troops and observers, usually with the consent of the parties involved, and a prerequisite was that a truce or cease-fire be in place; and the peacekeeping force was deployed primarily to observe that truce.

The international political forces that suppressed ethnic conflict and other rivalries have really, in many ways, been removed from the world. We're facing a growing number of these smaller conflicts, with the United Nations embarking on these peace missions with broader objectives, more demands, more players, and more complex mandates. In the words of General MacKenzie, a former Canadian commander of UN troops in Bosnia, "These new missions require a much closer integration of political and military assets such as civil authorities, human rights officials, electoral monitors, humanitarian assistance organizations, and civilian observers as well as military peace-keeping forces."

The United States and UN Peacekeeping

Over the last several months, the   U.S. military has become increasingly involved in support of UN peacekeeping, and humanitarian assistance operations toward these multinational and multilateral efforts are positive. But there is concern over some new dimensions that we are beginning to witness.

There were some thirteen UN peacekeeping operations in the first forty-three years of the UN's existence. In the last four years, another thirteen have been launched. Totally, these operations have involved a half-million soldiers from a great number of nations; unfortunately, 800 of those soldiers have lost their lives. From these experiences, we have a significant body of knowledge. We also have somewhat of a precedent about peacekeeping operations.

If you look in Chapter VI of the UN Charter, there is not a word about military forces or military units associated with peacekeeping. These forces were not mentioned at the time of drafting the UN Charter; it has never been amended, and it may never be. Military involvement has been an initiative of the United Nations, so peacekeeping operations can be considered "extra- charter."

We have learned many lessons from these experiences, especially in regard to peacekeeping. The United States has played a limited role up to now because there had been a perception within the UN that it would be difficult for officers from the Security Council's permanent five members to be impartial observers. However, U.S. military participation in UN operations has recently grown a great deal.

The United States has committed more than 500 service members to the UN. We have about 130 involved in observer duty in Angola, Cambodia, Iraq, Kuwait, Middle East, Western Sierra and Mozambique. We also have more than 400 soldiers and civilians with the UN in Yugoslavia, including staffers at a sixty-bed hospital. It's the first time we've ever put a unit secunded to the UN.

Three Paradigms

There are three different views of UN peacekeeping that are relevant for the U.S. military; these three views can be considered, to borrow a term from scholars, paradigms.

The first paradigm is really peacekeeping, or what we have called *truce keeping*. In these operations, military organizations may be required to observe and monitor terms of a truce or serve as an early warning for potential breaches of the peace. They rely less on the size and capability of their unit and more on the willingness of the belligerents, or former belligerents, to adhere to protocols. The protection of the force comes not from its own capabilities but from its neutrality, or perceived neutrality, in the conflict. A couple of examples are the UN Observer Mission in El Salvador (ONUSAL) and the UN Mission for the Referendum in Western Sahara (MINERSO). Another good example outside of the UN umbrella is the contingent the United States has in the Sinai as a result of the Camp David Accords. We have 1,000 soldiers there who patrol the line between Egypt and Israel. We have been patrolling this area since about 1979 and it's a pure example of peacekeeping.

The second paragigm (and it has been given many different names) is called *conflict management*. It contains many of the new peace operations that the United Nations has undertaken in the last couple of years. You might refer to these as "Chapter VI-plus" operations. This dimension is the new one referred to earlier, about which we in the U.S. military have some concerns. These operations are more complex and greater in scope than traditional peacekeeping, but they fall short of Chapter VII—in which you might go in search of a total victory.

Some suggest that the UN take on more of these missions. But as we develop four subcategories of conflict management, one can see that we're easing into something different and more difficult. The first subcategory is blockades of supplies to countries. We did it in the Persian Gulf and now we're doing it in the Balkans. In the second subcategory, we deny one or more parties the ability to conduct movement—a no-fly zone, for example. These zones are in place over Iraq and Bosnia. The third subcategory is protection of humanitarian relief operations. This effort is going on now in the UN Protection Force (UNPROFOR) in the Balkans and in Somalia. A fourth subcategory is military assistance to civil authorities. This subcategory would include restoration of civilian authority after a conflict has been resolved, much like what occurred in Kuwait after Iraq was expelled.

The final paradigm, and one of the more troublesome aspects from a military prospective, is *internal conflict resolution* which requires interposing UN forces between belligerents. These efforts are troublesome because such an operation puts neutrality at risk. These situations will be clearly problematic for the soldiers of the force involved.

At the risk of oversimplification, on one hand there is peacekeeping as described in relation to Chapter VI; on the other hand there is Chapter VII peace enforcement, similar to that deployed in the Persian Gulf. But the Persian Gulf campaign was not "pure" Chapter VII. Pure Chapter VII is total UN control; in the Persian Gulf, there were UN resolutions with, for the most part, U.S. control. The Chairman of the Joint Chiefs of Staff, General Colin Powell, has clearly stated, regarding Chapter VII, that when the national command authority determines U.S. interests are at risk and military force is deemed appropriate, Americans can be sure that their armed forces will be ready, willing, and able to accomplish the mission.

Concerns over Conflict Management

Our concerns today are over those operations associated with the paradigm termed *conflict management*. These military operations are complex and require a very responsive command and control setup, detailed planning and significant logistic support. The UN just does not have the requisite planning capacity to do these things. The planning has to come from a member state, which is a difficult enterprise.

As we examine potential U.S. involvement in these "Chapter VI-plus" conflict management operations, several issues arise associated with the commitment of U.S. service members to United Nations. General Powell also said, "Decisive means and results are always preferred." In conflict management operations, rules of engagement usually require minimum use of force. While it may be interesting intellectually to differentiate between limited action and war, when soldiers are on the ground getting shot at, the one thought in their mind is "I'm in combat and I want to use everything at my disposal to suppress that fire and end any injuries to my unit." There is a greater chance of having higher U.S. casualty rates in limited operations than in operations in which decisive use of force can be employed.

Another concern arises when combat operations are used to accomplish these missions. The real danger is that a UN force will be perceived as having taken sides. Imagine taking sides if there is peace enforcement intervention in former Yugoslavia. There are several sides to take. Once the intervening force takes sides, the primary form of protection—neutrality—is lost. One must only think back to the Marine Corps barracks in Beirut ten years ago to see what happens when neutrality is lost.

Still another concern with limited operations, and again quoting from General Powell, is that "sometimes perception [exists] that policymakers do not care whether we achieve victory or not." If commanders must take another step and gradually escalate, there evolves a situation in which one must escalate—step by step—more and more. History has not been kind to those commanders who escalate gradually.

Sometimes all situations may not be clear and one may not know when military force is the right answer. But military force cannot be used imprecisely, nor can it be used out of

frustration over not being able to resolve a conflict diplomatical-
ly. Sometimes, it just might make the conflict worse.

Another lesson learned from the past concerns preparing
the American people to know that they're committing their sons
and daughters to a possible protracted situation. If this preparation is not done, there may be a real problem with popular
support when casualties start increasing and the American public
has not been prepared politically for this eventuality.

The final concern relates to units that deploy for peacekeeping. Within the UN mandate, the sort of weaponry units
they may take is usually proscribed. The British and other
nations in UNPROFOR are light infantry. If the conflict
escalates or one of the belligerents has more weapons, we have
to ensure that we're not undermatched.

Conclusion

The United States military clearly understands that
peacekeeping and peace operations are important missions for our
nation. We have supported them in the past and we'll continue
to support the growing commitment the United States is making
to the United Nations to the extent that our national command
authorities direct us. However, we must remember that when
you use the word peace as a title, it's very benign. It is easy,
however, to slip from being peacekeepers to entering an armed
conflict without a conscious directive from the U.S. executive or
legislative branches, or the people of America.

UN Financing for Peacekeeping

I would like to begin with what some have referred to as
the UN's difficulties, particularly the rather weak infrastructure
for organizing peacekeeping and the scale of peacekeeping in
recent years and, indeed, months. I will concentrate on the
problems of paying for peacekeeping through the UN system, and

the solutions that have been proposed in the past year by both the current Secretary General and his predecessor. These solutions have, by and large, not been implemented.

The UN's capabilities for traditional peacekeeping and "new"peacekeeping verging into enforcement as "Chapter VI and three quarters" can be thought of as an accordion. If the UN is to respond quickly to particular emergency peacekeeping situations, it needs to expand quickly and have the capacity to pay for a rapidly expanding response. Thus, to pursue the accordion image: the UN organization must provide the very sturdy metal black frame that can hold the tough, flexible grey elastic in the middle together as it expands rapidly. The major troop contributors and the other major contributors that provide logistic and financial support for peacekeeping operations constitute the grey elastic. But you also need a kind of sturdy metal black frame on the end that can hold this gray stuff in the middle together as it expands rapidly. Both need to be strengthened and better coordinated.

As a result of the crisis over financing the 1956 UN Emergency Force One (UNEF 1) force in Suez and the 1960 Congo operation, it was determined in 1962 that the costs of peacekeeping are a proper and regular cost of the United Nations. The UN can therefore pay for them through mandatory assessments on all member states. Since 1973, every time a new peacekeeping operation has been approved by the Security Council, it has been paid for by a special assessment. Furthermore, since 1973, peacekeeping has been paid for on a scale of assessments different from the scale of assessments by which the regular budget of the UN is financed. The special assessment scale provides for a very substantial proportion of the total cost of peacekeeping to be paid for by the five Permanent Members of the Security Council, with a substantially reduced cost borne by the less-developed countries. Currently, 56 percent of all peacekeeping costs are borne by the five Permanent Members of the Security Council, with a substantially reduced cost borne by the less developed countries.

Six Problems

There are six problems associated with financing peacekeeping. These problems and some of the solutions that have been proposed are enumerated below..

First, the scale was just increased very rapidly in 1992. The costs of peacekeeping are projected to be approximately $2 billion, which is four times as great as they've ever been in the past (the previous peak year was 1989).

The second problem, an obvious one, is the emergency nature of peacekeeping operations. These emergencies are coming rapidly now, requiring the establishment of two or three or four new operations every year since 1988. The emergency calls for peacekeeping contributions arise at unpredictable times, which makes it very difficult both for UN budget planning and for national government budgets to respond to the call for an assessment payment.

The third problem is that there has arisen in the last twenty years, since the scale of assessments for peacekeeping was more or less established, some real anomalies in who pays for what. Currently, the greatest problem is that the former Soviet Union, which is expected to pay a rather large share of peace-keeping—12 percent of the total—is seriously in arrears because of its economic problems. But in recent years, the United States has also been seriously in arrears. The two countries together account for more than 60 percent of the arrears payments owed the UN for peacekeeping operations.

The fourth problem is the effect of the ongoing financial crisis of the regular UN budget on peacekeeping. The regular UN budget has been kept at zero real growth since 1986, which is a virtual requirement of the U.S. government and several other major donors. Yet this constraint keeps the peacekeeping operations department at fourteen or fifteen full time staff supervising the operations of 53,000 troops. It is this kind of constraint on the regular budget that is a drag on the system, no matter how rapidly a particular peacekeeping operation expands.

The fifth problem is that, given the real financial crisis of the UN as a whole, its reserves have been periodically depleted completely. So when the UN wants to respond to an emergency situation, it simply doesn't have any reserve funding to do so, nor does the Secretary General have authority to do so. In 1973, for example, UN Emergency Force Two (UNEF II) could be deployed within forty-eight hours to the Suez Canal in a real crisis in the middle of the Yom Kipper War. That mobilization would be inconceivable now because there is simply no financial reserve base to do it.

Finally, there is a pressure, particularly in these "Chapter VI and three quarters" missions that border on enforcement, to turn to countries for voluntary funding of peacekeeping/peace enforcement operations rather than to go the mandatory assessment route. In the minds of many, this pressure seems to erode the collective responsibility for international peacekeeping that has been an assumption of the UN for the last twenty years.

Some Solutions

Former Secretary General Javier Perez de Cuellar and current Secretary General Boutros-Ghali have suggested in the past year a number of approaches to solving these problems. The solutions basically address the problems of unpredictability, given that sometimes very large forces need to be raised in very short periods of time to accomplish very complex functions. The proposals have tried to address two needs: (1) the need to quickly access financial reserves to pay for the beginning of an operation until the General Assembly can agree to a budget and (2) the need to legitimize the burdens that various countries share. Both

secretaries general have endorsed, more or less, the following list
of solutions:

■   To permit the Secretary General, in an emergency
situation, to suspend competitive bidding.
■   To give the Secretariat advance approval of a portion
of a peacekeeping budget until the General Assembly has
approved the whole thing.
■   To provide for much more substantial UN stockpiling
of material and equipment.
■   To provide—instead of eleven different assessments
for eleven different peacekeeping operations that are distributed
throughout the year with eleven different letters to treasury
departments—for one single payment. This change would allow
for an overall peacekeeping budget that recognizes peacekeeping
as it has become the core activity for the United Nations. (There
would a line for each particular operation, but the entire budget
could be drawn upon to start a new one.)
■   To revise the peacekeeping assessment scale to get rid
of some of the anomalies and make sharing the burden seem
more equitable. (There is, needless to say, a big argument about
equity being different in the eyes of different beholders.)
■ To adjust various national policies so the accordion can
move out very rapidly when needed. (In fact, the United States
has led the way in this effort by passing the International
Peacekeeping Act of 1992, which for the first time set up a
contingency fund to finance the variety of peacekeeping bills that
would come due during the year. This fund is something very
few other countries have and we have set a very good example,
but there are other examples. The Nordic countries have had, for
some years, standby units and regular training programs that
other countries could attend. The Japanese have standby production lines for UN vehicles, both Nissan and Toyota. There are
various ways in which nations could provide for the rapid
expansion of UN capabilities.)

■ To provide for enhanced financial reserves of the United Nations, including a general reserve from which both the peacekeeping and regular budgets could draw, and the establishment of a particular reserve for peacekeeping.

■ To establish a peace endowment fund that would not depend on government contributions, but would solicit funds from other sources in the international system: from foundations, such as wealthy individuals, international financial institutions, and governments that wanted to donate money.

A Muted Response

All these proposals have met with a rather muted response. Possibly the only one that will carry the day, in at least these first years of Boutros-Ghali's tenure, is the establishment of a small peacekeeping reserve fund, with funds being drawn from surpluses in existing peacekeeping budgets, primarily the Namibia and Iran/Iraq border force budgets.

The reason is not really money. In this year of quadruple peacekeeping assessments, the 1992 cost of UN peacekeeping is almost precisely the size of the 1992 budget of the New York City Police Department. So the scale of financing for all the nations of the world is not an insurmountable sum. It is rather a problem of political control.

Many of the major donor governments, but particularly the United States, have absolutely insisted that the United Nations not have autonomous reserves and autonomous control over even very small amounts of money to be used as the Secretary General and the Secretariat wish. One has to conclude with this very basic question: Having described the range of the tasks and opportunities that lie ahead for UN peacekeeping, is it in the interest of the United States to provide the UN system with minimal independent capacity to fund some of these operations in their start-up phases? This is one that the United States and other governments have not been willing, at least so far, to address.

The United States and UN Capabilities

I am a constructive, but impatient and critical, observer of the United Nations and U.S. policy toward it. Ever since the term "new world order" has been used, there has been some satisfaction at the UN that it is a focus for differing crises in a way it never has been. With the realization of how radical the UN's powers actually are under Chapter VII—and that once the UN gets into something it has a habit of never getting out—countries have become concerned, certainly, about empowering the Security Council too much. This concern is buttressed by a concern that the UN is being used by the United States. This feeling is shared by many countries friendly to the United States, as well as others.

UN, U.S. Woes

There's also a tendency, when the United States goes to the UN, to deal only with part of a problem. This tendency is certainly evident in Somalia and in the former Yugoslavia, where we have regarded the problems as humanitarian in nature rather than political, which would require simultaneous solutions to all those aspects as well. We have taken this approach for good reasons, because it is easier, perhaps, to focus on the tragedy of ordinary civilians than it is to get involved in the political messes that some of these leaders have created.

The Secretary General, however, in his options put forward on Somalia, has very tightly connected these two parts of the problem. As this operation goes forward it is going to be very, very difficult to deal with only the humanitarian issues; it becomes obvious that there is another side. Sometimes, resolving the political side of the equation has to involve the use of force.

Another problem is the image of the United Nations. Certainly, in the United States, the image prevails of an organization that is a massively wasteful, lumbering, and frequently incapable bureaucracy, as described by the *Washington Post* in a series of articles appearing last October. Until there is a willingness at the UN to reform itself, and that willingness in not there now, there will be a general reluctance to let the UN be in charge of the serious business of war, if you will, even if peacekeeping itself has generally a more positive image.

There is a willingness in the United States to use the UN, but also a great reluctance to be bound by it. We don't pay our regular dues or peacekeeping assessments. This fact has to be stressed because there is a willingness to get engaged in other UN-sanctioned operations from the Gulf War to Somalia, where the question of financing doesn't come up because the United States, obviously, will foot the bill for whatever it contributes to the operation. A real willingness to pay what are legal obligations, not just favors to the United States, is lacking. This fact is very important.

The Role of Sovereignty

It is important to realize that the United States. is very interested in retaining its own sovereignty, as it talks about the sovereignty of other countries, despite our Security Council veto. At the same time the United States is not willing to empower some of the organs of the UN that would take away some of our own sovereignty, from supporting the Military Staff Committee to signing Article 43 agreements (which would assign standby troops to the UN which could be used, of course, only on Security Council authorization where the United States has a veto).

What strikes me the most is not how involved the UN is in peacekeeping, but how much the United States is identified not with peacekeeping, of course, but with enforcement. That is how the world sees the United States: unwilling to be identified with peacekeeping in the present, despite some steps that have been taken in the past few months, but quite willing to be identified with enforcement. I am struck by the small numbers of Americans involved, not by the trend lines.

When analyzing some of the actions in which the UN has been involved, some people say that Iraq's aggression against Kuwait was the easy case because it was aggression. In some ways, Somalia is the easy case because it is a TV tragedy, an effect that some people have called the "CNN [Cable News Network] phenomenon"—meant to be something that people have a consensus about. Even the Africans are convinced, or partially convinced, that something has to be done. Just two weeks ago, when there was another crisis and some refugees were leaving Somalia, General Aideed (a leader of one of the major Somali factions) said, when the United States criticized some of his actions, that there should be no foreign interference. He didn't want any. He made this statement as the refugee workers were saying they needed safer conditions.

It will not be easy to intervene and get out. The issue of sovereignty is perhaps not as great an issue in Somalia as it will be in other cases, because there was no state; there was basically chaos. One must ask: If there had been a humanitarian crisis, even of these proportions, and there had been a state in power, would the international community have engaged as it has in Somalia? I do not think the answer is necessarily yes. We've watched people die in many places of the world, from Cambodia on, and done nothing.

Part of the reason that force is acceptable is that ultimately the goal is the re-establishment of sovereignty. Certainly, that is the goal in Somalia, and it is the goal in Bosnia, or should be. If something is not done, there soon won't be any Bosnia to reestablish.

The U.S. Military Role

Turning to the Defense Department role, there is a willingness to come in for a military part of an operation: secure airfields, distribution centers, and so on. My question is: Then what? For example, what happens if the United Nations wants to establish a trusteeship in Somalia? Some people have floated this idea and said there's chaos, there's really nothing that can be done. Will American troops stay? Will they be American troops or will they then be transferred to UN command? What will happen, in other words, when we start looking at this problem as a political one and not just as a humanitarian crisis? Will we just say we've done our job and then go home?

The Defense Department and the UN, to put it mildly, have separate cultures in the way they act. The UN is, obviously, concerned mainly with words, political considerations, incrementalism, and resolutions that are sometimes passed in order to do something and sometimes in order to avoid doing something. The Defense Department, I would like to think, mostly tries to do something. Of course, the military would prefer to have clear missions and would prefer to have clear objectives. I think that kind of advice, while very well spoken, is like the advice of your broker when you ask what to do on the stock market and your broker says, "Buy low and sell high." That advice is very good, and if one could follow it, it would be very nice. The question is, Does the world allow you to always follow that advice?

The United Nations Association (UNA) recently sponsored a mission to investigate peacekeeping around the world. UNA sent about twenty individuals from all walks of life, including retired U.S. military officers, to Cambodia, El Salvador, Cyprus, and the Golan Heights. It was clear that the conditions acceptable to other countries participating in peacekeeping would not be acceptable to Americans; the differences ranged from medical facilities to command and control to many other things, such as communication. The questions really are: Can the United States, when it gets involved in a peacekeeping operation or an enforcement operations, always ensure that the conditions it needs will be there? Can the United States always ensure that the mandate it receives will be as clear as it likes? The answer is no and perhaps we'll have to deal with that kind of ambiguity.

Turning to Somalia, for example, it is good that the international community is doing something, but unfortunately the UN's role in enforcement is delayed. The current operation puts aside questions of command and control, the role of the Military Staff Committee, and questions of a standing UN force. It delays the question that should be examined: What is the role of the United States in enforcement under the United Nations, and is the United States ready to negotiate Article 43 agreements, as the Russians are? Russia has a bill before its parliament that would authorize it to sign Article 43 agreements that place troops under UN command, and France has said it would be willing to do that, too.

Look at the options the Secretary General laid out to the Security Council for action in Somalia. One of them, of course, is to accept the U.S. offer, which it did. The one he clearly preferred, the fifth option, was to have the UN take the lead. What would happen if the UN were to say "thanks but no thanks" to the American offer and "We don't want Americans to come in, we want this to be a UN operation?" Then the question of whether the United States is more concerned about control or compassion would have to be addressed. I am not sure of how that question would be answered.

The United States has created a situation in which it won't pay for UN operations unless there is U.S. control. So there is no viable alternative to a U.S. operation; in effect, the UN has no choice. Thus, the UN accepted the U.S. offer in Somalia.

The question of the mandate, and who determines it, is muddled and will remain so. There will not be absolute clarity about when force is to be used; what to do will not be clear. There are a lot of questions that are not raised and, certainly, not being answered. I particularly emphasize that I don't think what is happening in Somalia is relevant to the situation in Bosnia-Herzegovina, where the level of force needed, the time frame, and certainly the European desire to get involved are quite different.

Conclusion

Until the United States makes peacekeeping and peace enforcement a more integral part of its Defense Department mission (which may mean, perhaps, funding peacekeeping from Defense Department monies) and much more integral to American military thinking, the U.S. response will always be *ad hoc* and late. It will always be an American operation and not a UN operation.

Also, until the UN begins to develop some capabilities of its own (which, of course, means that the United States and all the other permanent powers on the Security Council would have to agree to such capabilities), the UN response is also always going to be ad hoc and late. It is revealing when people say "The UN does not *have* the capability," when what is really meant is the United States and the other major powers will not *give* the UN the capability." The United Nations is not an institution that *takes* the capability; it's one that's *given* the capability. To put it in the former way is to hide and confuse the issue. It makes no sense because the UN doesn't have capabilities; its member states do. If the United States and the other countries don't give it the capabilities, it will never have the capability to activate itself. Until the UN has something in place, nobody's going to look at it seriously. And it won't have anything in place, until the United States takes the lead to put it there.

Supporting Democracy

Elizabeth H. Ondaatje
Moderator and Rapporteur

The sessions of the national defense University symposium on non-traditional missions for the U.S. military progressed along an imaginary spectrum of unconventional activities from those that are relatively close to the Army's primary combat missions (e.g., regional deterrence, containment, peacekeeping, peace enforcement, and humanitarian intervention) to those that are far from combat or the threat of combat (nation assistance, security assistance). The subject of the session summarized in this paper, "Supporting Democracy," resides on that more benign end of the spectrum and involves the military in such activities as helping other nations develop and maintain public infrastructures and train and equip their militaries. The first two panelists outlined the key policy instruments of the U.S. military in supporting democracies abroad: nation assistance and security assistance. The second two panelists placed those instruments in the internal, interagency context of U.S. foreign assistance and in the external context from the perspective of the host nation receiving U.S. military assistance. The more general discussion allowed the panelists to expound upon these initial

Elizabeth H. Ondaatje is an International Policy Analyst, RAND since 1987. She does research and writing on Army noncombat activities, conventional arms control in Europe, Air Force planning and programming, and terrorism in the United States. Ms Ondaatje holds memberships on the Council on Foreign Relations; International Institute for Strategic Studies; and The Royal United Services Institute.

angles and provide insights into the trends, lessons, and prospective new policies that will affect U.S. military support for foreign democracies.

This paper summarizes the individual presentations made by four panelists. First, Lieutenant General Teddy Allen, Director of the Defense Security Assistance Agency discusses the military security assistance portion of the Foreign Operations Appropriations Bill as a tool for foreign policy. Next, Colonel Antonio J. Ramos, Director, Policy and Plans, United States Southern Command, continues the discussion on security assistance programs when he provides information on nation assistance programs in Southern Command. Reginald J. Brown, Assistant Administrator for the Near East, U.S. Agency for International Development, then shares his views on the problems and areas for consideration in coordinating interagency cooperation in U.S. programs which provide support to democracy. Lastly, Brigadier General David A.B. Mark, Director, Nigerian National War College, discusses the role of the African military in nations which are transitioning to democracy, and explains why the military must first take charge, and once in charge, find it difficult to step down. The paper closes with some conclusions and recommendations regarding the role of the United States military in supporting democracy.

Military Security Assistance

Military security assistance or security assistance is one of three parts of the Foreign Operations Appropriations Bill for foreign aid each year. The other parts are economic support and peacekeeping. As a tool of foreign policy, security assistance is a State Department program, but the Department of Defense implements it.

Security assistance is made up of two programs: Foreign Military Sales and International Military Education and Training.

Foreign Military Sales (FMS)

This multi-billion dollar program includes both sales of military equipment and financing for the purchase of military equipment by other countries. Currently there are $80 billion worth of goods and services in the pipeline. In 1991 Foreign Military Sales peaked at $23 billion, $12 billion of which was related to Operation Desert Storm. The impact of this program on the domestic economy is 50,000 direct and indirect man-years in the U.S. defense industry for every $1 billion in foreign military sales. However, due to the sense in Congress that military assistance is not as important as it once was and the pressures of the budget deficit, Congress reduced the program from $23 billion to $15 billion in 1992 and 1993. A closer look at the budget reveals that of the $23 billion in sales in 1991, $19 billion came from "cash paying customers" such as Saudi Arabia, Japan, and Korea. The remaining $4 billion in the program is set aside for grants and loans to some twenty countries. In 1992, Congress earmarked 84 percent of that $4 billion for six countries. In 1993, $4 billion was reduced to $3.3 billion, 97 percent of which Congress earmarked for six countries. Two of those six countries receive $3.1 billion (Israel and Egypt). This leaves about $89 million for other parts of the world (including fighting the drug war in Latin America) where the President, the State Department, and the Defense Department want to influence policy.

International Military Education and Training (IMET)

This $40-$50 million program enrolls foreign military officers in U.S. military schools such as Fort Leavenworth, the National Defense University, and the War Colleges. In 1991, Congress authorized $47 million for 125 countries; in 1992, $44.5 million; in 1993, $42.5 million. However, in 1992 and 1993, Congress also authorized sixteen more countries for IMET, including Czechoslovakia, Poland, Hungary, the Baltics, Russia, Ukraine, and several other former Soviet republics. This investment pays off ten to fifteen years later when these officers have risen to positions of importance in their governments and militaries. Many believe it helped account for the ability of the U.S. to forge a coalition in Operation Desert Shield/Storm among thirty countries with common doctrine, tactics, equipment, logistics and communications. In 1991 Congress also authorized an expanded IMET program (IMET/E) to allow the U.S. military to train nonmilitary governmental officials with oversight of the military. In 1993, DoD will begin training legislators, primarily in the former Soviet Union and Eastern Europe, with responsibility for military matters, primarily in the resource management area.

Not only are the resources for these two DoD security assistance programs shrinking while demand is expanding, but DoD has also received additional authorities to apply this diminishing pool of foreign aid each year to biodiversity, counternarcotics, POW/MIA efforts, and nonproliferation. At some point in the near future a reassessment of the goals, objectives and resources for foreign assistance, both military and economic, is needed and may be requested by Congress. DoD's program should be part of that reevaluation process.

The Defense Department's objectives for these programs are discussed by answering the following question: *What does DoD seek to achieve with security assistance?* Since the end of the Cold War, the Defense Department has had three objectives in mind when recommending to the State Department, Congress and the President how to allocate military security assistance money: first are base rights (including overflight rights and landing rights, not just stationing); next is Middle East peace; and finally the counterdrug war in South America. However, once the process reaches Congress, lobbying from interest groups frequently overrides DoD and State Department recommendations. As an example, the U.S. had been providing grant aid to

Turkey, Greece, and Portugal for years until Congress eliminated all grant aid and reduced the total available for loans to these allies by 10 percent in 1993. Consequently, DoD has to find new ways to maintain base rights in NATO's southern tier.[1]

Nation Assistance

Nation assistance is a continuation of the security assistance programs previously discussed. As security assistance budgets become tighter, nation assistance assumes greater importance in influencing other countries.

In this section we will outline the strategic rationale behind the Southern Command's nation assistance program, describe the program itself and detail its implementation, and finally, discuss the impact of nation assistance in Latin America.

Strategic rationale for nation assistance

In Southern Command, nation assistance is a means to an end, not an end in itself. The end is host nation capability to strengthen and maintain democratic institutions. This is one of the theater strategic objectives in the Southern Theater Strategy.[2] Nation assistance is not a separate strategic objective of the command; it is part of all the objectives. It is one of several means (along with counterdrug operations, counterinsurgency support, and military professional training) at the disposal of CINCSOUTH to achieve theater objectives and must be integrated with those other programs. The key idea is to develop host nation capabilities to provide infrastructure and institutions to meet citizens' expectations. Fulfilling citizens' expectations, it is argued, can ensure the legitimacy of the government.

Defining nation assistance

The principles underlying the definition of nation assistance are:

- Sustainability
- Sound economic base
- Environmentally sound development
- Develop host nation capabilities
- U.S.-only is the last resort
- Credit the host nation
- By host nation request only

It should be emphasized that nation assistance can only be provided if the host nation requests it through the U.S. Ambassador. Applying those principles, the Joint Staff has developed the following working definition of nation assistance:

> Political, economic, informational, and military efforts to support host nation development. The main objective in nation assistance is to assist the host nation in developing self-sustaining institutions. Military nation assistance programs that support U.S. policy objectives and host nation development activities are an integral part of the U.S. Government's country plan or strategy. The military component of nation assistance includes: engineering, medical, civil affairs, and other military resources that contribute to infrastructure required for institutional development in the host nation.

In other words, nation assistance includes more than military activities and is more than construction projects. It is part of an overall political, economic, and informational effort. It is seen as a way to build common values and ideals in the American hemisphere for the 21st century. The focus of nation assistance is on building host nation capabilities (i.e., nation building by the host nation supported by SOUTHCOM), rather than U.S. military capabilities. Training is not the primary objective though it is an obvious military benefit given that the CINC aims to use the same tactics, techniques, and doctrine that they use at the operational level in any other kind of operation. The mission is stated thus:

> USSOUTHCOM assists nations in the Southern Theater by supporting host nation development of self-sustaining capabilities for nation building; supporting host nation development of responsive institutions of government; and, by providing engineer, medical, psyop, civil affairs, and other military support and training responsive to the host nation's needs.

Implementing nation assistance in SOUTHCOM

SOUTHCOM has planned and conducted nation assistance activities for more than twenty years. The CAPSTONE operational plan describes the ways, means, and ends of the nation assistance program. It is SOUTHCOM's peacetime engagement operational plan. The keys to the program are the U.S. Ambassador and Country Teams. Together they develop the country plan; SOUTHCOM develops operational plans to support it. As with the theater strategic objectives, the major operational programs in the plan are mutually supporting. Nation assistance programs are designed to work with, rather than separately from, host nation efforts. Competing with private road building or other construction interests is usually not an issue since the basic infrastructure does not exist. In the majority of the Southern Theater countries, there are vast areas the government does not control. In these places the U.S. military assists the government in developing a positive presence by building roads and other infrastructure. The U.S. military may interact with more than just the host nation military; their interlocutors may be the Ministries of Agriculture or Commerce. Nation assistance is not a free program to the host nations: the U.S. supplies the manpower, equipment, and know-how, but the host nation purchases supplies.

Nation assistance planning and operations can be categorized as "deliberate," "surge operation," and "crisis action." Deliberate planning and operations span two fiscal years, are integrated with the theater strategy, and are based on the CINC's priorities.   Surge operations support host nation counterdrug operations, span a 90- to 120-day time frame, and are executed during Stage IV of the surge operation.   Crisis actions are essentially disaster relief planning and execution.

In 1992, sixteen of the nineteen countries in the SOUTHCOM area of responsibility received some kind of nation assistance from the U.S. military. Of the 25,000 individuals who deployed to the Southern Theater in 1992, 18,000 deployed as part of the nation assistance program. In 1993, SOUTHCOM expects 13,000 individuals to deploy to the Southern Theater for nation assistance, possibly including all nineteen countries. The forces which conduct nation assistance activities are not the combat divisions, wings, and battlegroups that the other CINCs plan for. USCINCSO's forces of choice are engineers, doctors, civil affairs specialists, lawyers, and Special Operations forces from all the services.

Actual activities include the following:

- Exercises
- Deployments for Training (DFTs) and Military Training Teams (MTTs)
- Foreign Military Financing Program (FMFP)
- Psychological operations and humanitarian, civic assistance programs
- International Military Education and Training (IMET)
- Medical Readiness Exercises (MEDRETEs) and Dental Readiness Exercises (DENTRETEs)
- Subject Matter Expert Exchanges (SMEEs)
- Engineer activities
- DoD Excess Property program

- Command and Control Support
- Civil affairs seminars including civil affairs doctrine and organization, civil-military cooperation, civil-military regional planning, and disaster relief

Nation assistance in SOUTHCOM also includes expert exchanges in public affairs, legal and human rights training, reserve force development, training and doctrine development and military tactics. In 1992, SOUTHCOM's nation assistance program built 110 kilometers of roads, 10 bridges, 88 schools, 40 clinics; dug 14 wells; held 17 civil affairs seminars; and treated 91,000 medical cases and 61,000 veterinary cases. For 1993, SOUTHCOM plans for 200 engineering exercises, 130 medical exercises, and 26 civil affairs seminars.

What is the impact of nation assistance?

Nation assistance impacts on democratic development in Latin America. It is uncertain whether or not the undemocratic behavior of the military in Venezuela and Argentina is an anomaly. Although there are still some elements in many Latin American militaries that refuse to accept the military's role in a democracy, these militaries have taken quantum leaps in changing their mindsets. For example, in El Salvador the military accepted civilian control, reduced forces, eliminated certain leaders, and played a role in the peace plan. In Venezuela, it was the military that put down the rebellion by what a Venezuelan questioner in the audience described as a small fraction of the military who do not have a clear understanding of the role of the military in a democracy. And in Brazil, the military stayed in their barracks during the impeachment process. It is difficult to change the military mindset in countries where it has controlled the government for over twenty years or more. The thrust of SOUTHCOM's efforts is to assist these militaries in understanding the role of the military in a democracy and to make the military part of the solutions to the problems in that region rather than part of the problems.

In conclusion, it should be emphasized that nation assistance is not the only part of the Southern Theater strategy, but it is a necessary part. Nation assistance is one of chief operational methods of achieving theater strategic objectives. Nation assistance is a means to strengthen host nation democracies and build the legitimacy of host nation governments.

Interagency Considerations

This section of the paper will provide background on USAID aired at a mainly defense-oriented audience, suggest several causes of the current problems in interagency coordination, offer a few possible solutions, and comment on the specific question of coordinating U.S. support to democracy.

Background on USAID

USAID operates under the same legislative authorities as security assistance programs. Within the total foreign assistance program, USAID either manages or directs developmental, economic, and food aid. Table 1 summarizes the various components of foreign assistance and USAID's role within them.

TABLE 1: Foreign Assistance Programs

Program	Develop-mental Aid	Economic Support Fund	Food Aid	Multilateral Aid	Military Assistance
Directs Policy	USAID	State	USDA	Treasury	State
Admin-isters Programs	State	USAID	USAID	State	DoD
Requested budget	$2.9B	$3.5B	$1.3B	$1.9B	$4.2B
Percent of total budget	22%	27%	10%	14%	32%
Actual budget	$2.5B	$2.7B	?	?	$3.3B

What this table shows is that there are at least five federal agencies with significant resources available to support nation building and economic development overseas, not to mention the Congress. USAID, for example, deals closely with four Congressional committees. This observation raises the question of how best to coordinate the efforts of these various agencies.

The problem of interagency coordination

Two primary challenges for coordinating complex interagency activities are coordinating policies and programs in Washington, and coordinating the implementation of those policies and programs in the host nation.

With regard to the first challenge, the problems of in-country coordination are rooted in the fact that each element is an extension of an independent agency or department in Washington. These agencies and departments have separate legislative authorities that created and fund them. They each have distinctly different expertise, assets, and interests. Consequently, coordination is difficult at best. For example, USAID leadership was unaware of plans for Operation Just Cause in Panama until it was launched. Yet USAID was expected to send disaster assistance teams to assist with post-war reconstruction, without notice or prior planning. This sort of rapid response is difficult for an agency such as USAID that operates programs under very specific legislative authorities with tight budgets under close scrutiny.   USAID does not have discretionary money for unplanned activities; it cannot switch money from one account to another without prior notification. In the end, the question of what would be done and who would pay were resolved, but not without   a   great   deal   of   rancor   and   accusations   of uncooperativeness and unresponsiveness against USAID.

As for the second challenge of coordinating implementation, the U.S. Ambassador is responsible for the overall coordination of the in-country activities of all U.S. Government agencies. In addition to the State Department, the agencies usually represented at U.S. missions in developing countries are the Departments of Commerce, Agriculture, and Defense, and the United States Information Service and the Agency for International Development (the latter two are, of course, part of the State Department). Compounding the problem of coordinating among so many actors is the fact that agencies duplicate efforts. For example,   USAID has conducted some of the nation assistance activities previously discussed in more than eighty countries for many years. Nevertheless, because there is someone in charge in-country, namely the Country Team under the direction of the U.S. Ambassador, there is a focal point for direction and guidance. In SOUTHCOM countries, this coordination works quite well. However, back in Washington the agencies do not coordinate well because no one is in charge.

A third challenge relates to defense intelligence requirements and economic aid programs. The need to ensure that

USAID activities are not confused or combined with intelligence gathering requirements is paramount. Mixing the two objectives will lead to problems.

What are the possible solutions?

The solution to the problem of interagency coordination in Washington is a nemesis. Nevertheless, some suggestions toward a partial solution follow: First, the lesson learned from the Operation Just Cause experience (as well as from Operation Urgent Fury in Grenada) is that there is a desperate need for interagency cooperation whether engaged in crises or in the ongoing business of supporting democracy. The Defense Department must involve USAID in the planning of an operation if DoD intends to tap USAID resources. The defense establishment must first recognize that agencies, such as USAID, have a role to play in this area and that better interagency coordination is required. However, it will always be difficult to orchestrate all the various participants, both public and private, from Washington.

A second possible solution involves the Policy Coordinating Committees run by the Assistant Secretaries of State to coordinate interagency activities in Washington. However, these Committees work well when dealing with specific problems; they are less effective for coordinating sustained operations in a country and cannot compare in effectiveness with the Country Teams. Yet, working outside the PCC framework is difficult. A recent effort to set up an Inter-Agency Task Force on Third World stability led by the Assistant Secretary of Defense for Special Operations met with State Department opposition. The State Department objected to dealing with this long-term issue outside of the PCCs and the Task Force was abandoned.

Ultimately the National Security Council is the body to address problems and propose appropriate solutions for interagency coordination.

Supporting democracy

Among the organizations and agencies that play a role in American support to democracies around the world are the Bureau of Human Rights, the Regional Bureaus, and the U.S. Information Agency at the Department of State; the quasi-governmental National Endowment for Democracy and the related Democracy arms of the Republican and Democratic parties, the U.S. Chamber of Commerce, the AFL-CIO, and a host of other nongovernmental organizations; and, of course, USAID. The rationale behind USAID's and other assistance efforts is that the prosperity and security of the United States is best sustained and enhanced in a community of nations which respect individual, civil and economic rights and freedoms. USAID provides $100 million each year to support the further- ance of democracy in nearly 100 countries around the world through its various programs (e.g., improving governance and accountability by identifying mechanisms to make officials more accountable to their constituents and to help constituents under- stand their rights and responsibilities as consumers of public goods and services; legal and judicial reform; election reform and monitoring; improving revenue generation and internal budgeting and auditing; and improving municipal services such as water and sewerage). However, supporting democracy is a foreign policy question and, therefore, must be controlled by the State Depart- ment. USAID's approach has been to work with State and with embassies in design and execution of those programs that support democracy and democratic institution building. But, again, coordinating from Washington will be difficult, particularly with efforts to support democracy. That is why the Secretary of State gave USAID the mission of supporting the world-wide trend towards democracy. USAID takes this mission seriously and is trying to execute it through programming in the embassies. USAID faces a dilemma: On the one hand, USAID and others cannot effectively perform these activities without putting more energy into effective interagency coordination as well as coordination between the Executive Branch and Congress. But because this is so difficult, the real emphasis or focal point for USAID activities will most probably be at the Country Team level under the direction of the Ambassador.

The Role of the Military
in a Transitioning Democracy

This section of the paper will describe the economic and political challenges facing struggling democracies in Africa, explain why the military in Africa finds it easy to step in and take charge, yet once in charge, finds it difficult to step down, and conclude with several recommendations for U.S. policy in Africa.

What are the critical problems in Africa?

The most critical problem is loss of aid and support. African countries are the real losers of the Cold War. The West won the Cold War by stopping the spread of communism. But the Warsaw Pact countries also won because the Western countries are supporting them and finding a place for them in the community of democratic nations. African countries that were wooed during the Cold War by the West and/or the Warsaw Pact countries are now are completely forgotten, despite the fact that many regimes in Africa are ripe for transformation into democratically elected governments.

The next problem is declining exports. Most African countries are completely dependent upon one agricultural product. But these agricultural products are not in high demand and in some cases are now supplied by other developing countries. For example, Latin American countries are replacing Africa as suppliers of cocoa to North America, because they are close, and the United States is focusing on encouraging the peaceful development of the Western Hemisphere. Other exports, such as granite, are being replaced by synthetics. And Africa will lose to other agricultural exporters in the General Agreement on Tariffs and Trade (GATT), because Africa does not have the wherewithal to meet the requirements of GATT.

Third, there is a lack of infrastructure. African countries lack social and technological infrastructures and those that do exist, do not function regularly. This in an area in which Africans are in desperate need of assistance from the U.S. military through the aforementioned nation assistance programs.

The fourth problem is corruption. Developed countries often decline to give aid to African countries because the leaders are corrupt. But it is possible that the problem is not as severe as perceived: The total budget of all the governments in sub-Saharan Africa is far less than the total amount required for the American savings and loan problem. Nevertheless, there is often a mismanagement of natural and financial resources which African countries must remedy.

Fifth, population growth and inadequate health care are significant contributions to African problems. Both of these critical problems are a function of the lack of functional education in Africa. AIDS is just one of many diseases to which Africans are succumbing. Malaria is also a very common disease. Unless African countries provide both functional education and a health care system, AIDS and other diseases will continue to spread. It should be noted that Congress has strongly supported population and AIDS programs in developing countries. Congress earmarked $350 million for population programs alone, which is more than the Bush Administration had requested, and mandated that an Office of Population be set up within USAID.

And finally, in addition to all of these problems, civil wars have generated millions of refugees. According to one estimate, Africa has over 60 percent of the world's refugees, and this statistic may be underestimating the problem.

This is far from an exhaustive list of Africa's problems. What is important to take away from it, is the recognition that if the West allows the situation in Africa to continue to deteriorate, and if Africa becomes more remote from the West, such a policy of neglect will not enhance the long term interests of the West. Observers predicted years ago that one day the Cold War would be over and the rest of the world would have to worry about the exportation of poverty from Africa. We may have reached that day.

Before turning to the question of what the U.S. and the U.S. military in particular can do, a discussion on the temptation for the military to take charge and the difficulty in developing democratic institutions and practices once they do would be educational.

Why does the military often step in?

Given this situation, the military often steps in to "take the bull by the horns" because the military is still the most cohesive body in most African countries; it represents the national symbol of patriotism; it is usually the one organization that tries to remove itself from ethnic problems; and young college graduates with degrees in political science and other social sciences join the military and do not have enough to do in the service to occupy themselves. Without enough to occupy their minds ("an idle mind is the devil's workshop"), they examine national issues, not from the perspective of the military supporting the government, but as critics of the government. They begin to think that if the government can not solve problems, maybe they should try.

Therefore, after watching the situation deteriorate, the military often moves in and seizes power. Once they are in power, they realize the country's problems are beyond their control (e.g., debt from military acquisition during the Cold War when the Soviets sold the African nations hardware, knowing they could not repay; nevertheless, the debt still appears in World Bank records). Moreover, they have more to do than they can handle because they still have to look after the military (both civilian and military components after a coup) as well as run the country. And yet, once in power, it is difficult to leave; therefore, there is another coup. But before the new group settles down to governing, it is time for another coup. The resulting instability obviously keeps foreign investment at bay.

Why is it difficult to replace military rule with democratic rule?

There are two major challenges in the transition to democratic rule. First, officers who are exposed to political positions once they are given a political appointment find it difficult to return to the barracks and simple military lifestyles. Political office becomes their next career target. The solution that the Nigerian military has attempted is to require that officers leave the military for good if they decide to go into political office. In addition, they have tried to keep mid-level military officers as far away from the spoils of political office as possible. The higher-level officers that are exposed are already senior enough to think about leaving military to go into politics anyway.

Second, it is difficult to develop a legitimate electoral system. Every time Nigeria has had a military change of government, it has been traced to corrupt elections that bring an unpopular leader into power. Consequently, the current government in Nigeria rescheduled January's primary for August this year after reports of corruption. They wanted to ensure that the process was legitimate so that no one would have an excuse to overthrow a popularly elected leader. However, this move was interpreted by some as the military postponing its departure from power.

What should the United States do?

Six suggestions for U.S. policy makers on what the United States should do in Africa are offered in the following paragraphs:

First, do not abandon Africa now that the Cold War is over. A world in which the gap between the "haves" and the "have nots" and which makes "a nuisance and nonsense" of the United Nations is not a good world. Developing countries look to the United States and the West for assistance, particularly as they make the difficult transition to a democratic form of government.

Second, policy makers must recognize that there are many forms of democracy and many paths to achieving democratic rule. It is difficult to evenly balance the demands of democracy, economic development and national stability. From the perspective of a two hundred-year old democracy, it is often difficult to appreciate the challenges of a developing country moving toward democracy. And from the perspective of a citizen of a developing country, it is difficult to await the outcome of the confusion patiently. The military is particularly impatient. The military thinks it will be fulfilling its responsibility to the country by taking care of the chaos. Given these pressures, it may be necessary to define democracy in Africa uniquely.

Third, beware of sending mixed signals to emerging democracies. For example, it is difficult for other African countries to understand why, despite Mobutu's dictatorial, undemocratic and suppressive policies, the U.S. still gave assistance to Zaire. Kuwait is another example of sending the wrong signals to developing countries. Africans believe that the U.S. could have asked for an interim government and eventually a democratically-elected government, rather than restored the Emir. As a world leader, the U.S. must send clear and consistent signals.

Fourth, do not let the media's focus dictate U.S. policy

priorities. The information that the media captures and transmits back to the U.S. has a powerful impact on U.S. policy priorities. These images have a lasting effect and often obscure what is going on off camera. The situation in Somalia is no worse than the refugee situation in Liberia or Sudan, but received U.S. military assistance as a result of the media coverage. The U.S. should be concerned about what U.S. troops would do there after the initial food distribution.

Fifth, provide more training opportunities for African militaries and civilian government officials alike. To help achieve democratic elections and assist the civilian government to remain in power once elected, Nigerian military and civilian officials need training. Of the total amount of U.S. training assistance, very little goes to Africa. Most goes to countries in the Western Hemisphere and the Middle East. African countries need training in civil affairs, medical care (health care is worse in Africa than Asia), drug interdiction (there is a lot of drug trafficking in Nigerian, though drugs are not grown there), and enhancing and maintaining regional stability.

Sixth, support regional security organizations. The Economic Community of West African States (ECOWAS) have banded together in a cease-fire monitoring group (ECOMOG) to address the Liberian problem. Initially only Nigeria, Senegal, and Guinea contributed. Now only Nigeria and Senegal fund it because the others cannot afford their contributions. Consequently, Nigeria finds itself in a leadership position in the West African sub region, but is eager to obtain support and assistance from the West. As the de facto leader in the region, it is essential that Nigeria's transition to civilian leadership progress well. Nigeria will be in an impossible situation if it assumes leadership in the West African sub region and then does not adhere to democratic principles. Nigeria will find itself isolated and deprived of any possible benefits from the world community for its efforts on behalf of West African regional stability. The U.S. and UN must also strengthen the Organization for African

Unity if it expects this weak organization to solve regional problems or to be a regional partner in situations such as Somalia. Recently the UN and the O.A.U. have worked together on Somalia and that when Ambassador Oakley flew to Africa to work on reconciliation, he worked with the O.A.U., and the O.A.U.'s Secretary General, in particular. Furthermore, the UN has conducted a parallel diplomatic effort with the West African nations. Regional security organizations might play a regional coordinating role among the Country Teams that coordinate nation assistance in various host countries.

Conclusions and Recommendations

Several key points emerge about the role of the U.S. military in supporting democracy.

First, the debate over whether or not military assistance, nation assistance, humanitarian assistance, and the like are appropriate roles for the U.S. military is polarized between those who see these as important U.S. military means to enhance security in the post-Cold War world of nontraditional threats and regional instability and those who fear that these activities will recklessly divert the U.S. military away from their single-minded focus on combat readiness for the inevitable next war. Both sides agree that the end of the Cold War presents the U.S. with new challenges (e.g., stemming weapons proliferation, employing traditional military forces against nontraditional threats) and new opportunities (e.g., shaping military reform in the former Soviet republics, creating and/or enhancing regional stability), but disagree on the extent to which these are appropriate challenges and opportunities for the military to address.

Second, within these polarized discussions, there was very little research and analysis provided to support either sides' contentions and concerns. For example, what is the actual (versus anecdotal) impact of nation assistance on the countries of Central America? What leverage or goodwill does the U.S.

actually earn with military assistance and what are the pitfalls? On the other side of the coin, how can the positive training benefits of nation assistance or the positive impact of the IMET program be quantified? Anecdotally it is credited with part of the success of coalition warfare in Operation Desert Storm. What was its actual impact and what are the implications of reducing the IMET program for future coalition contingencies? And, finally, what are the risks associated with not adapting the military to the nontraditional requirements of the post-Cold War world?

Third, there is a shrinking pool of money in the DoD budget with which to encourage and support emerging democracies. What little remains is either earmarked or spread so thinly over so many countries that there is little more that security assistance programs will be able to do in the future to support democracy overseas.

Fourth, regional organizations are playing and will continue to play an increasingly important role in resolving regional conflicts (in conjunction with the United Nations) and encouraging democratic development.

In turn, these four key points logically offer a series of recommendations. First, as security assistance budgets tighten, nation assistance assumes greater importance in supporting democracy. The integrated program of nation assistance, counterdrug, and other noncombat activities in SOUTHCOM may provide a model to other CINCs to achieve their theater strategic objectives.

Second, given budget constraints at home and the potential for spreading democracy abroad, it is time for a rational reassessment of U.S. foreign aid to determine if it is properly allocated and if it is having the maximum impact. As a consequence of this reassessment, the U.S. should release the foreign aid budget from the grip of Middle East policy and scale back foreign assistance commitments to Israel and Egypt, in particular. That money would then be made available to emerging democracies in the Eastern Europe and the former Soviet Union.

Third, a related recommendation would be to try to limit the domestic political considerations that sometimes steer money away from important regional allies, such as Turkey.

Fourth, rather than call these activities "nontraditional" since the military does have a long tradition of performing them, or "nonmilitary" since the military does perform them, call them "noncombat." This term ensures that they are not perceived as any less a traditional, military responsibility than warfighting. Having made this distinction between combat and noncombat, there is at least one qualification to make. Namely, some activities, such as humanitarian assistance or drug interdiction, may fall into a "gray area" between combat and noncombat because, unlike a program such as nation assistance, they can quickly escalate into combat. This gray area has been conceptually outlined by a former CINC as "aggregated roles" of combined noncombat and combat operations and "rapidly shifting roles" from noncombat to combat operations and back again. Developing a common vocabulary to describe the activities and concepts of this symposium is an important first step in bridging the communications gap that exists between the two sides of the debate on the future of U.S. military noncombat activities.

Notes

1. Turkey is particularly important to the U.S. as it seeks to encourage pro-Western (or at least not anti-Western) policies among Turkey's Islamic neighbors. The Turks provide a positive example of the combination of a democratic and an Islamic state. The U.S. seeks Turkey's help with the southern tier states of the former Soviet Union. Of course, the history in this region complicates all possible solutions since at one time or another almost every country was a victim of some other country. The only hope is to look at each country in the regional context, not in isolation.
2. The other theater strategic objectives are: assist host nation in eliminating threats to regional security; support continued economic and

social progress; assist host nations in defeating drug production and trafficking; with Government of Panama, ensure open and neutral Panama Canal; enhance military professionalism.

Supporting the Civil Authorities at Home and Abroad

Seth Cropsey, Moderator
John R. Brinkerhoff, Rapporteur

"The armed services today have to be versed not alone in war but in government, politics, the humanities—economics, social, and spiritual."

Bernard Baruch[1]

THE ISSUE OF THE ROLE OF MILITARY FORCES in support of civil authorities in the United States or in other nations focuses on whether the Department of Defense and the Armed Forces ought to be engaged in these non-traditional (non-combat) missions. This paper will examine both sides of this issue, and present four examples of current military support to civil authorities that illustrate what is being done today. Examples of current operations in support of civil authorities

Seth Cropsey joined The Heritage Foundation as director of its Asian Studies Center in January 1991. Prior to his current position, Mr. Cropsey served as Principal Deputy Assistant Secretary of Defense for Special Operations and Low Intensity Conflict. He also served as Assistant Editor of *The Public Interest*, and reported for *Fortune* magazine on U.S. private enterprise and public policy.

John R. Brinkerhoff is a consultant on national security matters with special interest in mobilization emergency management, civil military operations, and the reserve components. He served on active duty in the United States Army retiring in grade of Colonel. He is also a graduate of the Army War College.

include Department of Defense humanitarian assistance programs; the California National Guard in the Los Angeles Riots; Coast Guard maritime law enforcement operations; and involvement of military personnel in domestic natural and technological disasters.

There is general agreement that the term "non-combat" is better than the term "non-traditional" because many of these kinds of missions have been performed in the past and are currently being performed by the U.S. Armed Forces, as shown by the examples to be discussed. The real issue is not what is or is not traditional, but whether military forces should perform non-combat missions.

One side of the issue is the position that it was not appropriate to use military forces for missions other than those involving combat, while the other side holds that the use of military forces on non-combat missions is not only appropriate, but highly desirable. The divergence between the two sides grows smaller as the discussion of differences of opinion focuses on three specific topics: the raison d'etre for military forces; impact of non-combat missions on readiness of military forces; and benefits accruing from such use.

The Case Against Non-Traditional Missions

The purpose of the Armed Forces is to "win our wars." It would be undesirable to change the mission of the U.S. Armed Forces from combat to disaster relief or some other humanitarian function. Military forces are to be formed, organized, staffed, equipped, and trained for fighting in combat operations, and any diversion of purpose to non-combat missions is undesirable. If these kinds of non-combat operations became the mission—or even the partial mission—of the Armed Forces, the troops would lose the edge from their training for combat, and would be less likely to prevail in war.

The use of military forces for non-combat missions could detract from readiness for combat. The skills needed to feed hungry people in Africa or build tent cities for victims of hurricanes are different from the skills needed for combat. Employment on or even training for humanitarian assistance missions will dull the fighting edge of troops whose mission is to kill.

Major changes in the utilization of the Armed Forces also could weaken support for combat oriented spending. In the present climate of fiscal austerity, diversion of military resources to non-combat missions would strengthen the hand of those who want to reduce further the Defense Budget. In the words of one discussant: If this thinking takes hold, will Congress take away the weapons from the 82nd Airborne Division?" One vessel could not simultaneously perform with equal proficiency the functions of a hospital ship and an aircraft carrier. There might be a tendency to protect support units--engineer, transportation, and supply--that are good for non-combat missions as the forces are reduced in future years. This could lead to Armed Forces that are under strength in combat units.

Moreover, there is no evidence that the people want the purpose of the Armed Forces changed from the combat role. While there is some superficial attractiveness to the notion of "trimming sails to the current winds" it is better not to be caught short if the wind changes. The use of civilian agencies and funding may be appropriate to achieve humanitarian ends, but using the Armed Forces is not.

The Case For Non-Traditional Missions

Combat is indeed the principal reason for having armed forces, but using the Armed Forces for non-combat activities makes good sense. Not only is something useful being done, but forces already in being (and already paid for from war preparation funds) are applied to do work that usually is in effect an unprogrammed workload. A force structure built for combat can respond to disasters also.

Combat oriented training of the military units and personnel is a good fit for humanitarian assistance missions. For example, military pilots trained for combat operations adapted well to the risks and challenges of flying relief missions in mountainous areas of Northern Iraq. Moreover, many "non-combat" missions (such as providing humanitarian assistance in Somalia) require some security, and that can be provided best by combat units.

The employment of military forces does not have to degrade readiness for combat. For a combat unit, prolonged participation in a non-combat role might dull readiness for combat, but for an engineer unit that would be the best training for the combat role. Non-combat operations are good training experiences for support units, since their non-combat missions parallel the roles they would perform in combat operations. Engineer units will build facilities, transportation units will carry equipment, people, and supplies. Medical units will care for sick and wounded people. The command and control system will work very much as it would for a war. Logistical operations will differ only in the kinds of supplies being distributed. Employment on civic action projects in Central America provided Army engineer troops valuable training that was applied later in the Persian Gulf. The stress of some non-combat operations is similar to that of combat, and some civil-military operations take place in a dangerous environment. Military operations during Hurricane Andrew, Hurricane Hugo, the Loma Prieta Earthquake, and Operation Provide Comfort were stressful because of long hours and difficult working conditions, but these kinds of experiences prepare troops for combat.

The Armed Forces should look to the American people for missions. The Cold War is over, and Desert Storm has already been forgotten. The Armed Forces have to remain relevant to the people and do what the people think is necessary, for without popular support they will not be able to fight and win future wars. If there is support for humanitarian assistance, disaster relief, civic action, and nation assistance as part of the overall United States response to a world in trouble, then the Armed Forces should be used to carry out the national will. It does not make sense to require the civilian agencies of the Government—already overwhelmed by current demands and neither organized nor equipped to perform these missions—to do the job alone while the Armed Forces sit on the sidelines waiting for war.

The question is whether nations that want and need help from the United States are going to be permitted the application of the full national capability for assistance. If not, the question about using the Armed Forces is moot, but if so, it would not make sense to forego the tremendous capability of the Armed Forces. In fact, the talent and resources of the Department of Defense and the Armed Forces are already being used extensively in non-combat roles that are perceived as in our national interests.

Department of Defense Humanitarian Assistance Programs

The Department of Defense (DoD) already has several large programs to support civil authorities abroad, generally called humanitarian assistance, but including also foreign disaster assistance, civic action, and peacekeeping. Military personnel and civilian employees of DoD work with foreign governments to provide resources and services. The current versions of these programs originated in the 1980s (but with ample precedents in prior years) in efforts to increase operations in support of democracy in Central America. The programs are planned, staffed, and implemented by the Office of the Secretary of Defense (OSD), the Joint Staff, the Military Services, and Transportation Command. Most of the programs are overseen by the Deputy Assistant Secretary of Defense for Global Affairs in the Office of the Assistant Secretary of the Defense for International Security Affairs. DoD Programs include Excess Property;

Transportation; Humanitarian Assistance; and Civic Action. DoD also supports the Foreign Disaster Assistance Program operated by the Agency for International Development (AID).

Humanitarian Assistance

Humanitarian Assistance is provided in the form of supplies, equipment, and troop units. Deliveries of humanitarian assistance have increased dramatically in recent years. In 1992, deliveries were about $15 million, and in 1993 will be about $25 million. This is a popular program with Congress. In 1992, Congress appropriated $100 million for transporting humanitarian supplies to the new republics of the former Soviet Union. Thus far, 2,500 tons of emergency food and medicine have been sent from the United States, 22,000 tons of food from Europe, and additional food from the United States. This program also involved stationing a military hospital in Tblisi, Georgian Republic, to provide medical assistance.

Excess Property

Excess property of the DoD is a source of non-lethal equipment and supplies for delivery without charge to nations in need of assistance. Excess equipment availability is made known to the Department of State, and equipment requested by nations is turned over by DoD to the Country Team for distribution.

Transportation

Transportation by military aircraft and ships is available both on a space available basis and a dedicated basis. Much of the excess equipment provided to other nations is delivered by military transportation assets. Military aircraft have been used also to medevac wounded Afghans to Germany and the United States for treatment in civilian hospitals. Space available transportation is also available to carry privately donated humanitarian supplies and equipment to nations in need.

Title 10 Humanitarian and Civic Assistance

Title 10 Humanitarian and Civic Assistance is administered by the Regional CINCs directly, with approval authority and administrative responsibilities vested in OSD. This authority permits U.S. military forces on training exercises or deployments overseas to conduct humanitarian or civic assistance projects with DoD paying the costs of the materials. U.S. military forces have used this authority throughout the developing world to build hundreds of schools, clinics, roads, community centers, bridges, and water systems. Numerous medical readiness exercises have been conducted by military doctors, nurses, and medical technicians. During FY 1991, medical and engineering humanitarian or civic assistance projects were conducted in 30 countries at a total cost of $3.5 million.

Foreign Disaster Assistance

Foreign Disaster Assistance is supported by DoD in the form of transportation and occasionally material and personnel. DoD responds to requests for assistance for the USAID Office of Foreign Disaster Assistance (OFDA). These requests, authorized by the Economy Act, are validated by OSD and then tasked to the Joint Staff. OFDA frequently provides funding for DoD emergency transportation. The Regional CINCs have authority to act unilaterally during the lifesaving phase of natural or manmade disasters overseas, such as the eruption of Mount Pinatubo in the Philippines. The FY 1992 Defense Appropriations Act provided $25 million in funding to be used by the CINCs to cover unanticipated costs incurred in foreign disaster assistance. Major disaster assistance operations conducted in the recent past include: Operation PROVIDE COMFORT in Northern

Iraq on behalf of the Kurds at a cost to DoD of $450 million, and Operation SEA ANGEL in Bangladesh, which cost $20 million.

The California National Guard In
The Los Angeles Riots

Active and Reserve Component units and individuals of the Armed Forces are used to assist in maintaining law and order during civil disorders in the United States. The National Guards of the several states are used frequently in this role as state troops under the command of their respective governors. Most of the time, the presence of the National Guard to augment local and state police forces is sufficient, but in some cases additional Federal military forces also are needed to accomplish the mission and restore order. The most recent major civil disorder involving the Armed Forces was the Los Angeles Riots in April 1992 following the jury verdict acquitting the police officers accused of beating Rodney King.

The 40th Infantry Division (Mechanized) of the California National Guard was one of the major military organizations participating in the quelling of this civil disorder. While the jury verdict in the Rodney King beating trial was a surprise, the 40th Division commander and staff had anticipated that some civil disturbances would likely follow such a finding. Although the 40th Division had been informed officially that they "would not be needed," actions were taken to speed up a response if the Division were needed. These actions paid off when the call came.

The Division was ordered at 2130 hours, 29 April 1992, to assemble 2,000 troops at their armories for civil disorder duty. Shortly thereafter, the Division was tasked to provide another 2,000 troops. Eventually, 80 percent of the soldiers of the 40th Division were called up as well as the entire 49th Military Police Brigade. A total of 6,000 National Guardsmen served during the riot.

The initial element of troops was ready to go into action at 0400, 30 April 1992, only 6 hours after the order was received. This was a fast response, and overall the Division put more people on the street faster than at any earlier time. One reason for the rapid response was that the order came late in the evening when most personnel were at home and could be notified easily.

The area affected by the riot was large, about four times larger than the Watts riots of 1965. It spread from the Ventura Freeway in the North to San Pedro in the South and from San Pedro to Pasadena in the East. Most of the intense activity, however, was concentrated in the West Los Angeles area.

The first step was to establish command and control of the Division units and effect liaison with the local authorities. The two major law enforcement agencies involved in the riot were the Los Angeles County Sheriff's Department and the Los Angeles Police Department. There are 87 cities in Los Angeles County, and 44 contract with the Sheriff's Department for law enforcement services, so it is a major player in the area. The prevailing lack of cooperation between the Sheriff's Department and LAPD proved to be a disadvantage in trying to bring forces to bear on the riots in a coordinated manner. The Division set up its own Emergency Operations Center at the Armed Forces Reserve Center at Los Alamitos and sent liaison teams to the EOCs of the Sheriff's Department, LAPD, and Long Beach Police Department.

While the troops were assembled waiting for orders to deploy to the riot locations, they received refresher training on riot control operations, fire discipline, and the rules of engagement. This training paid off. The Division demonstrated excellent fire discipline. Although there were thousands of soldiers on duty carrying rifles with live ammunition, only 21 shots were fired, and all of these were determined to have been justified by the rules of engagement. The Division carried out its duties as it had been trained to do without major problems.

Tactical transportation was provided by local busses. As a mechanized division, the units are equipped with tanks and armored personnel carriers that were not needed during the riots and would have caused damage to the roads if they had been used. To replace the armored tracked vehicles, supplement the Division's organic trucks, and provide adequate tactical mobility, buses were obtained from the Rapid Transit District for Los Angeles County and the Orange County Transit Authority. The agencies were well organized and very cooperative, and they did a good job in carrying the troops to and from their operational locations. Air National Guard aircraft were also used to expedite the movement of some elements of the Division from their home stations in Central and Northern California.

The Division operated in support of the local police departments under the control of the State of California, which assigned tasks to the Division EOC. All missions for Division troops were planned and assigned by the Division Headquarters. As troops were shifted around to meet the needs of the situation, there were complaints by some local mayors that "their" National Guard units were being taken away from their cities and used elsewhere. It was necessary to explain firmly but politely that the National Guard was to be employed statewide (or in this case city-wide) to meet the needs of the situation rather than be reserved for the security of its home station city.

The conflict between the LAPD and Sheriff's Department caused some confusion and inefficiency. Aid that could have been provided from the County was not well received nor utilized by the LAPD, and police from the California Highway Patrol were assembled and available but not used well.

On May 1st, the 40th Division and other California National Guard troops on state active duty were called into Federal Active duty by the President. This was a surprise to the Division and the local authorities as well, for the decision had not been well coordinated locally before it was made. A Joint Task Force (JTF) to command the Federal Troops was formed in accordance with standard procedures. Major General Marvin L. Couvalt, Commander of the 7th Infantry Division at Fort Ord, California, was placed in charge of the JTF. The 40th Infantry Division Commander, Major General Daniel J. Hernandez, was appointed the Army Forces Commander for the operation, and the 1st Brigade of the 7th Infantry Division, was placed under General Hernandez during the operation. Federalization worked well operationally, and all of the military personnel worked together smoothly.

There are differences of opinion, however, about the necessity and desirability of calling the 40th Division and other California National Guard personnel to Federal active duty. The operational consequences were to modify the rules of engagement, cause problems with contractual actions already underway, and shift the financial burden from California to the Federal Government. The California rules of engagement allowed a soldier to insert a magazine into his weapon if told to do so by a police officer. (The soldiers were walking joint patrols with police officers.) The Federal rules of engagement were more cautious and did not permit local decisions on when to load with live ammunition. This became an emotional issue, but it was worked out after awhile. A more serious problem was taking care of actions to purchase support and services which were started as California contracts under state rules and then converted to Federal contracts under different rules. This change caused a lot of extra administrative burden but did not interfere with the provision of support. From the State of California's viewpoint, the assumption by the Federal Government of the cost of the military augmentation forces was a blessing and made Federalization a good thing.

However, some National Guard officials in California and Washington, DC, believe that the decision to Federalize the 40th Division was unnecessary and unwise. The officers and soldiers of the California National Guard were trained in supporting civil authorities in disasters and disorders and had experience in performing these missions. They had served often in these situations and believed that they knew what to do better than active Army forces, who do not often routinely support civil authorities. Moreover, the act of Federalization brought to bear on the troops a general DoD prohibition against exercising police powers pursuant to the Posse Comitatus Act, even though the law itself may not be as restrictive as the DoD policy. As state forces, the National Guard troops could be used as police or in close teamwork with police, but they could not do that as Federal forces. Overall, the National Guard believes that—except for the funding support—it would have worked just as well or better if the 40th Division had remained on state active duty and had worked in a cooperative arrangement with the Federal forces sent in to help.

As it was, the operation went well from the military perspective. Peacetime training paid off. Division headquarters benefitted from participation in Warfighter Exercises and Command Division Refresher Training; the state of individual training was high; and the use of the standard five paragraph field order at all levels of command helped simplify and speed up tactical decision making. The Division physical fitness also paid off--in one respect by helping the troops work long hours without degradation of their capability. The 3rd Brigade of the 40th Division drove all night from their home stations in Northern California and found that the original plan for them to stop for a 6 hour rest was canceled because they had to be in the riot zone urgently. As a result, the soldiers of this brigade went about 40 hours without sleep but managed to do their jobs.

A good feature of the operation was the response of the people of Los Angeles to the National Guard troops. The people welcomed the soldiers and supported them in their riot control operations. They brought so much food and snacks that the soldiers were "eating like kings." Soldiers were regarded as more objective than the police and were considered to be "honest brokers" capable of serving the people's interests.

Coast Guard Maritime Law
Enforcement Operations

The United States Coast Guard has a dual role in support of civil authority in the United States. It is by statute both an Armed Force and a Law Enforcement Agency. As an Armed Force it prepares and trains to fight as part of the Department of the Navy during wartime. As a law enforcement agency it is engaged every day as part of the Department of Transportation in an unceasing struggle against criminal activity. The Coast Guard is one of the Nation's largest law enforcement agencies and the primary maritime law enforcement agency. The Coast Guard has in fact functioned in the so-called "non-traditional" roles for the entire 202 years it has been in existence. The Coast Guard also exercises several non-law enforcement domestic responsibilities, including marine safety, boating safety, aids to navigation, waterways management, polar and domestic ice breaking, and search and rescue. The four major law enforcement programs for the Coast Guard are Protection of Living Marine Resources, Maritime Safety Regulation, Alien Migration, and Contraband Interdiction.

Protection of Living Marine Resources

Protection of Living Marine Resources is the new term for fisheries law enforcement and is designed to secure marine life habitats and enforce adherence to relevant international treaties and domestic regulations.

Maritime Safety Regulation

Maritime Safety Regulation enforcement is designed to assure that ships are equipped and operated in accordance with United States law. Coast Guard personnel inspect ships and cite offenders for safety violations.

Alien Migrant Interdiction

Alien Migrant Interdiction operations are intended to prevent illegal entry by aliens to the United States by ship or boat. One of the main operations in this function has been the effort to prevent Haitian economic migrants ineligible for entry from sailing to the United States. Many such migrants originate in Haiti, and the Coast Guard has intercepted and detained numerous boat loads of Haitian migrants and either returned them to Haiti or taken them to Guantanamo Bay. In FY 92, the Coast Guard handled over 37,000 Haitian refugees from 470 boats. The Coast Guard also picks up many Cubans fleeing in small boats and rafts. Cubans are usually admitted into the United States as political refugees. Coast Guard patrols take place throughout the Caribbean and in the Florida Straits. The Coast Guard also responds along the Pacific Coast to intercept illegal immigrants from Asia—mostly from the People's Republic of China.

Contraband Interdiction

Contraband Interdiction is a major law enforcement mission. Smuggling to avoid tariffs was the problem that led to the formation of the predecessors of the Coast Guard, and contraband interdiction remains a mission today. Much is heard about interdiction of illegal drugs--a major mission its own right. However, the Coast Guard continues to work with the US Customs Service and other agencies to intercept, search, and seize other contraband to prevent entry into the United Sates or to assure that proper duties are paid.

Drug Interdiction

Drug Interdiction is the largest single Coast Guard program, and it takes up 19% of the Coast Guard operations budget. The Coast Guard is the lead agency for maritime drug interdiction. The goal of the maritime drug interdiction program is to disrupt the flow of illegal drugs from Central and South America. The Coast Guard works closely with the Drug Enforcement Administration (DEA), DoD, and several other Federal and state agencies. Drug interdiction activities take place in four zones: Source Country; Departure Zone; Transit Zone; and the Arrival Zone.

Source Countries are those from which illegal drugs--primarily cocaine and marijuana--originate. Cocaine typically is made from coca plants grown in Peru and Bolivia and processed in Colombia. From Colombia the drugs move to numerous other nations that serve as intermediate bases from which the drugs are smuggled to the primary markets in the United States. The Coast Guard maintains a presence in some source countries to help local maritime law enforcement agencies learn to cope with the criminal elements fostered by the illegal drug trade. The Coast Guard provides law enforcement training in 31 countries in partnership with the DEA and other US and local agencies. There are always more requests for training than can be met. In addition, the Coast Guard has given over 20 patrol boats to source nations to increase their own capacity for maritime interdiction.

The Departure Zone lies offshore from source countries and is the area from which ships carrying illegal drugs emanate. The role of the Coast Guard in the Departure Zone is to provide Law Enforcement Detachments (LEDETs) on Navy ships to board vessels suspected of carrying drugs and make arrests if the vessels are U.S. or appropriate special arrangements have been made with the foreign flag state through diplomatic channels. The primary departure zone for drug interdiction is the Southern portion of the Caribbean Sea off the coast of South America. Since 1984, the Coast Guard has participated in 26,000 vessel boardings resulting in the seizure of 138,000 tons of cocaine and 10 million tons of marijuana. (The Coast Guard also maintains

LEDETs on Navy ships in the Adriatic Sea and Persian Gulf to provide boarding parties to enforce embargoes on war materials to Serbia and Iraq, respectively.)

The Transit Zone is the area off-shore from the United States through which illegal shipments pass en route to the maritime approaches to this country. This is primarily the Northern part of the Caribbean Sea. The Coast Guard works closely with other agencies to intercept and seize drug shipments before they are landed in the United States. The Coast Guard has close ties and working relationships with other nations in the Transit Zone in which the resources of several nations are pooled to provide more effective interdiction. One of the more successful of these local arrangements is among the Bahamas and the Turks and Caicos Islands, which provides for sharing information and resources to interdict drug traffic in and around these islands. Throughout the region the Coast Guard works with the naval forces of friendly foreign governments, whose navies are likely to resemble the Coast Guard more than they resemble the US Navy, and who are likely to view the Coast Guard as a non-threatening presence because of its peacetime missions.

The Arrival Zone extends outward from the shore of the United States for approximately 50 miles. In this zone the Coast Guard works with the Customs Service, DEA, FBI, and local authorities to coordinate the interception of drug shipments and the apprehension of drug traffickers.

The Coast Guard is a relatively small force of about 37,000 active duty personnel backed up by a Reserve Component of about 10,000 personnel. It operates about 200 ships—all small compared to Navy surface combatants—and about 200 aircraft. The structure, strength, and funding for the Coast Guard is keeping pace as demand for its peacetime missions increases. Continuous employment of the Coast Guard in its ordinary activities do not detract from the readiness of the organization to perform its combat missions in wartime. In fact, conducting dangerous and demanding operations daily contributes to readiness overall.

Armed Forces in Domestic Disaster Response

The Department of Defense has engaged in responses to domestic disasters for many years, and there are well established procedures and plans for doing this. The Secretary of Defense has designated the Secretary of the Army as Executive Agent for Military Assistance to Civil Authority (MACA), and the Director of Military Support (DOMS) on the Army Staff is the action agent for the Secretary of the Army for this program.

Essentially, there are two military chains of command. For combat operations, the chain goes from the President to the Secretary of Defense to the Unified and Specified Commanders. For domestic non-combat operations the chain goes from the President to the Secretary of Defense and thence to the Secretary of the Army and the Director of Military Support.

The Director of Military Support is a Major General who is also the Director of Operations, Readiness, and Mobilization in the Office of the Deputy Chief of Staff for Operations and Plans, Headquarters, Department of the Army. DOMS has a permanent staff of six personnel that is augmented in time of emergency with representatives of OSD, the Joint Staff, the Armed Forces, the National Guard Bureau, and other Federal Agencies.

The Director of Military Support directs DoD participation in non-combat operations in the United States (including Hawaii and Alaska) and Puerto Rico, US Virgin Islands, Guam, American Samoa, and the Northern Mariana Islands. The DOMS focus is entirely on domestic emergencies, and it does not get involved in training of foreign military personnel or other foreign support activities. In the event of an emergency requiring military units or resources, DOMS will task the appropriate Service or CINC to provide the support. For the Los Angeles

Riots, DOMS tasked US Forces Command to provide the troops that ultimately formed part of Joint Task Force Los Angeles. Standing missions coordinated and implemented by DOMS include the following (with year of authorization in parentheses):

- Civil Disturbance Operations (1968)
- Support to US Postal Service (1970)
- Domestic Disaster Relief Operations (1971)
- Combating Domestic Terrorism (1971)
- Military Assistance to Safety and Traffic (1973)
- Support to Immigration Emergencies (1991)

DOMS provides support to the following Federal Agencies:

- Federal Emergency Management Agency for major disasters.
- US Fire Service for major forest fires.
- Department of Justice for civil disturbances, immigration emergencies, and loans of equipment.
- Environmental Protection Agency for major chemical spills.
- Department of Energy for radiological emergencies and energy disruptions.
- US Postal Service for postal service disruptions.
- DOMS also provides medevac support directly to local communities.

Major emergencies for which DoD has supported Federal agencies include the following:

- Idaho Wild fires (1988-1992)
- Exxon Oil Spill (1989)
- Hurricane Hugo (1989)
- Loma Prieta Earthquake (1989

- Talladega, Alabama, Prison Riot (1991)
- Typhoon Omar on Guam (1992)
- Hurricane Iniki on Hawaii (1992)
- Hurricane Andrew in Florida (1992)

DoD is a signatory to the Federal Response Plan (FRP) promulgated by the Federal Emergency Management Agency (FEMA) to coordinate the Federal response to domestic emergencies. DOMS is the active participant for DoD in the FRP, and provides the DoD representative to the Catastrophic Disaster Response Group that assembles at FEMA Headquarters to monitor events and resolve policy and program issues during emergencies. DoD is the lead agency for two of the 12 Emergency Support Functions: Public Works and Engineering (ESF #3) and Urban Search and Rescue (ESF #9). DoD is a supporting agency for all of the other 10 Emergency Support Functions. The US Army Corps of Engineers heads up the Public Works and Engineering Function, and DOMS manages the Urban Search and Rescue Function.

DOMS designates one or more Defense Coordinating Officers (DCO) to work with FEMA's Federal Coordinating Officer (FCO) in the affected areas. The DCOs respond to the FCO, interface between military organizations and local, state, and Federal Agencies; request support as needed from other Emergency Support Function Managers, and receive and validate requests for support from other Emergency Support Function Managers. If the emergency is large enough, a Joint Task Force may be formed, as in Hurricane Andrew, and the JTF Commander then works in concert with the lead Federal agency through the DCO.

During Hurricane Andrew in Florida, from 24 August to 15 October 1992, DoD provided substantial disaster relief support, including 24,000 DoD personnel; 900,000 meals, several life support centers (tent camps); numerous radios and generators, and over $140 million in contractor support. The first support was a shipment of 100,000 MREs (Meals Ready to Eat) that arrived on 25 August, the day after Hurricane Andrew hit. Military forces were directed by the President on 27 August to provide additional disaster assistance, following a request by the Governor of Florida for Federal military assistance, and the first hot meals were served the following day to victims in the disaster area. One of the first military missions for Hurricane Andrew was to bring in heavy equipment to remove debris from roads so that other disaster assistance personnel could enter the stricken area.

DoD is a major player in domestic emergency management, providing massive assistance and support through the Federal Response Plan and also rapid response by local military commanders direct to communities requesting urgent assistance for local emergencies to save lives and property. Support of civil authorities in domestic disasters and civil disturbances has been a mission of the Army and DoD for a long time.

The Future for Non-Traditional Missions

The non-traditional, non-combat operations already being performed by the Department of Defense and the Armed Forces showed that this is and has been an accepted set of missions. Non-combat missions are not new. The Armed Forces have been doing them for over 200 years.

The most traditional of the non-combat missions has been military engineering in peace and war. Not only have military engineers helped to build our nation but they have also performed useful work in other nations. In some cases, this has paid off in war also. Participation by the US Army Corps of Engineers in long-term construction of over $14 billion in projects in Saudi Arabia paid off for Operation Desert Storm. The relationships and trust that was built between US and Saudi officials during the construction was as important as the airfields, ports, housing, and other facilities that supported the Armed Forces in the war with Iraq. These construction operations showed the Saudis that we could be trusted to do good work and that we would leave when the job was over.

There was general agreement that deterring and winning wars remain the primary reasons for having Armed Forces. If performing non-combat missions requires abandoning a capability for combat, it would be wrong for the Armed Forces to perform non-combat missions. But no one is advocating that. The Armed Forces have the organization, training, discipline, and equipment to do both missions, so it is not an "either-or" situation. The Armed Forces are a total team consisting of full-time military personnel, part-time military personnel, civilian employees, and contractors. All of these members of the team are available for non-combat missions as well as combat missions. The consensus of the group was that the force structure ought to be designed for the combat mission, and then utilized as possible for non-combat missions. Moreover, the ability of the Armed Forces to perform non-combat missions well is a by-product of their training for the more difficult mission of combat, and if the combat mission is replaced, the Armed Forces will lose not only their ability to fight but also their ability to provide humanitarian assistance. Therefore, maintaining a capability to provide humanitarian assistance or perform other non-combat operations must always be incidental to maintaining a capability to perform the combat mission.

The use of the Armed Forces in non-combat missions is still a long way from becoming accepted as commonplace based on the opinions of the panel members. Generally the panel was opposed to suggestions by Senator Nunn that the Armed Forces might become involved in drug-counseling, urban assistance, and other broad social programs to help America. The consensus was that these kinds of missions would put the Armed Forces in direct competition with other public and private organizations, something that has not been welcome in the past and currently is prohibited by law.[2]

The panelists thought that the purpose of the Armed Forces in the Post-Cold War Era must be viewed in light of a broader mission than simply "to fight the nation's wars." The new, broader mission might be "to carry out the nation's will and meet its needs"—particularly for promoting peace.

Peace is as important as war, and a new role for the Armed Forces is to promote peace. Peace is the desired state. If deterrence fails, the United States seeks decisive combat and then a return to peace. The United States has not been good at transitioning from war to peace, as recent experience in Just Cause and Desert Storm indicates, and it is necessary to learn how to promote peace. If the larger view is not adopted, the non-combat missions that serve peaceful purposes may fall through the cracks. Peaceful use of military power is a most elegant and appropriate use.

Notes

1. Inscription on a plaque on the wall of the Baruch Auditorium, Eisenhower Hall, Industrial College of the Armed Forces, Fort Lesley J. McNair, Washington DC.
2. The Public Works and Economic Development Act of 1965 prohibits direct competition by the Armed Forces.

Final Thoughts: Non-Traditional Roles for the U.S. Military

John R. Galvin

SOME AREAS ARE ESPECIALLY IMPORTANT TO the question of non-traditional missions, which have proliferated since the end of the Cold War. To start, it is helpful to put non-traditional missions into some kind of context.

First of all, they fit into a time when things are not so predictable. To take an example, one could really say today that there are *two* Russias. There is the Russia in which Yeltsin's dreams come true. When Yeltsin took over, he said that in 7 months he would turn around the economy. That was a striking thought, and it hasn't come about. And he has done pretty well so far, but he's got a lot of trouble ahead. So there is the *first* Russia in which Yeltsin succeeds, and his country goes down the trail to democracy, free enterprise, a market economy, respect for human rights, and all the things that he has talked about. Is that what's going to happen? Would anyone be willing to put some money on that over that the next year or so?

Then there is the other Russia in which the hardliners take over and Yeltsin doesn't last that long. There are also ways

---

John R. Galvin is the John M. Olin Distinguished Professor of National Security Studies at the United States Military Academy, West Point, New York. He assumed this position July 1, 1992 upon his retirement after 44 years of military service. Prior to his retirement, General Galvin was the Supreme Allied Commander, Europe and the Commander-in-Chief, United States European Command from 1987 to 1992.

to look at the rest of central and eastern Europe, indeed, at a lot of other places. There's China. One could say there are *two* Chinas, one which does what China is doing right now, which is shoveling the coal into its economic locomotive while holding back hard the brakes of political change, and where will that lead? The other China lets go of the brakes, and that is danger- ous too.

To return to the question of context, one can develop some kind of architecture for security and stability in a world which does not currently display much predictability, a world in which there is a lot of instability. A great part of that architec- ture is out there in the form of a much more effective United Nations. In fact, Secretary General Boutros Boutros-Ghali says more support is needed for the United Nations from the regional organizations. And one reason is that the United Nations over the past 4 years has deployed more forces than in its 43-year history. In 1992 this cost $750 million; in 1992 it will cost at least $3 billion to support deployed forces. That's one reason for considering regional organizations, and there are already some regional structures. There is the Conference on Security and Cooperation in Europe for the European region, and subregional structures like NATO and the Western European Union. Regional structures are lacking in some of the other areas. But there is the Rio Treaty structure in this hemisphere and there have been some tries at structures in the Pacific. It is possible to look at the latter some more.

If there is a combination of that kind of structure, it might work better than predicted half way through this century, because people are more willing to put faith in structure and maybe not only in structure, but in processes such as arms control and nonproliferation of weapons of mass destruction, and also in relationships among the nations of the world, communi- cation, and crisis management.

There is still a need for standing forces, but what I have described is the kind of context to look at in terms of security around the world in the immediate future—a world approaching a new century. And this has been the worst century in terms of people slaughtered in war, many of whom—more than half—were noncombatants. No one wants to see a 21st century similar to the first half of this century, but many of the reasons that the second half hasn't been so bad is that there has been structure, and maybe it can be made to work even better. That is a worthwhile objective.

Turning to non-traditional roles, I may surprise you by saying that one of the first ways to look at them is collectively, not as an individual nation, because that is not the future for the United States, doing many things unilaterally. In those terms it's easier to think about unilateral action than collective action. Looking at collective action, the first implication it has for non-traditional roles is subregionally, underneath some kind of an architecture as illustrated by an alliance like NATO.

NATO's future probably will embody non-traditional roles, at least to some degree. There isn't any need for a "Fulda Gap complex" or a line of foxholes stretching from Norway to Turkey, but there is need for some other NATO capabilities—command, control, and communications and intelligence structure, and an infrastructure, 28 percent of which has been paid for by the United States, and that means the bases, ports, pipelines, airfields, and long-range communications that exist across the great span of Europe because they have been put there by this country.

But more than the fact that there is the infrastructure, command and control, interoperability or standardization or rationalization, the NATO military forces have operated as teams for more than four decades. This is something that is even more important about NATO when it comes to questions of the non-traditional role, namely, political-military interface—he North Atlantic Council•and the fact that the council has not only itself as a representation of the 16 nations, but military committees, infrastructure committees, budget committees, and the like. It has a military standing force under it and also a way of directing that force. This standing force of two million troops with a backup force of five million has been controlled by a coalition of nations for 41 years in a political-military sense. It has been disciplined and subordinated and controlled by those nations, a very unusual thing in world history. Structure like that is needed, and NATO can be the example.

These non-traditional roles also suggest another concept, that is, the fact that aggregations of these roles will change very rapidly. A role may be traditional, such as deterrence or conflict, but quickly shift into peacekeeping, peace enforcement, humanitarian assistance, or another role, and these might change very rapidly from one to the other—back and forth. This might happen in Yugoslavia. It certainly happened with the rescue of the Kurds, where 450,000 people were blocked at the Turkish border, not able to go any farther in escaping from Saddam, not able to return, stuck on these 45° slopes of high mountains. The first mission was to get food to them because they were starving to death. Initially, food was dropped within 24 hours, but flying up canyons dropping blind with full power and full flaps on C-130s was an adventure. Once it was realized that more had to be known about where the food was being dropped and what was needed, a change in mission was requested. This was non-Clausewitzian. The objectives were being changed in the middle of the effort. But it worked.

People were put on the ground and reported, "It isn't a matter of dropping food. The Kurds are dying because of a lack of water, poor sanitation, and other conditions." There were cholera and plague indications in these groups. They had to move out of the mountains. So there was a need for yet another change of mission. To move them out of the mountains, land was needed. It had to be taken away from the Iraqis as firmly but gently as possible in the hope it wouldn't cause conflict. The Iraqi military had to move out of an area 60 km wide and a 100 km deep. At the same time they were being told not to hurt the

Kurds, even though Kurdish guerrillas were attacking the Iraqis, and sporadic conflict continued ensued. Peacekeeping, peace enforcement, humanitarian assistances—all were going on at once. Then it was noticed that some Kurds were exfiltrating the 60x100 km area and going back home. Iraqi outposts were letting them by. This led to a desirable outcome to the problem. There wouldn't be facilities which turned into PLO-style camps since the camps became way-stations and the Kurds kept on going. So the Iraqis were asked what would they do if the Kurds went through. If they didn't cause trouble, would they be allowed to pass? They answered, "yes."

These changes meant U.S. troops had a different mission each week almost—sometimes two or three missions going on at once. If that happens the first thing needed is a strong political-military interface where one can get the answers and mission changes as well as concurrence and consensus in terms of what must be done. To do that and to do any non-traditional role, that kind of structure is required—if non-traditional missions mean collective missions.

This doesn't rule out the possibility of U.S. unilateral action. The United States is unique in the sense that it conditions its relationships with others on certain ideals that go back to the Revolution, such as human rights. The relationship with China is conditioned; practically the U.S. has nothing to do with China because of Tiananmen Square. This is the way that Americans respond to things. When one asks what is meaningful to an American, he or she will say freedom, democracy, individual rights, self-determination, and free enterprise. Those are the things for which the Nation stands. While that leads away from the first point about the nontraditional, that is working with an organization within a structure, within an architecture of security and stability, with or without it, the points are equally applicable.

A major non-traditional task is nation building. Nation building does not have a good reputation in Latin America or in this country and that's unfortunate. A turnaround in the status of

nation building is needed. Nation building is not a new task.
The U.S. military has contributed immensely to the building of
this Nation, of the United States. The U.S. military opened the
West, built the railroads, supported the railroads, provided the
engineers for them, and built the Mississippi River system, which
it still controls today. It fixed and drained and dredged our ports
and is still doing that. In terms of environment the Nation's
military forces control 600 of the major lakes in this country
today. And many other things have been done; one could
include such undertakings as the Panama Canal. It is a proud
tradition of nation building which the military of today shares.
And in all that was done to build the West, there was no military
governor of any territory. We asked for no political clout, for at
no time in U.S. history did military people seek political control.
That is something to convey to Latin Americans. Support for
democracy in nation building is important to the United States.

Taking up the critics one hears that there was a school
for the dictators in Panama, and that all the dictators in Latin
America went to it. That's ridiculous, although it is true that
there were military people who got involved in coups in their
Latin American nations. Mistakes were made in the sense that
a lot of foreign military people learned only too well what it's
like to live in a true democracy and when they went back and
looked at the democracy in their own countries they became
frustrated because they couldn't get anything done to improve
conditions.

It is necessary to understand that it is not just a matter of
military people creating problems in Latin American countries in
this century. It is a political-military question. Sadly, that
political-military interface has not been there because the political
leadership has not been concerned.

Looking at nation building makes non-traditional roles
appear more complex. It isn't just a matter of having a nontraditional role and going out and providing some training or military
assistance. It's much more than that. The task must be put into

context. There are U.S. Navy Seabees, for example, who have done marvelous jobs all over the world, and especially in Africa and Latin America. There are Air Force medical teams that have performed miracles. But there is a larger context which requires considering how to do non-traditional roles in such a way that they are successful and helpful in the long term.

In the case of a lot of humanitarian efforts, a lot of disasters—the Mexico City earthquake, the Armero mudslide, the San Salvador earthquake, the Bolivia earthquake, the many problems of Africa, some in Europe, the massive rescue operation of the Kurds, the Bangladesh operation—the military had what the volunteer organizations did not. The Red Cross and others are marvelous organizations but they don't have the command and control and communications or the logistics and organizational capabilities of the military. For example, these agencies used to think that while the military might be helpful in the first day or two of a disaster, after that there is really no place for troops. But when the disasters grow as big as they have lately, the volunteer agencies have changed their minds. The kind of non-traditional efforts the military makes, even if they are not multinational, are efforts of a coalition in which many other organizations are involved.

Finally, what does the military need in order to carry out the non-traditional mission? More training? More equipment, added doctrine, different doctrine? Different organization? Non-traditional roles are really crisis response roles. The role is always taken in a crisis when one is responding. It is fine to call a role non-traditional, but one also ought to talk about crisis response.

The military is taught to respond to crises, to make decisions when all the facts aren't there. This is what all the military schools teach, to take action under pressure, to work as a team, and to be troubleshooters. They teach the military to organize, reorganize, establish task forces, and do task reorganization and tailoring. So in many ways the military is already prepared, no matter what the organization--Army, Navy, Air Force, Marine Corps, Coast Guard. Some additional training may be needed, but one shouldn't get hung up on the idea that somehow a whole new force is needed to do these things. The military should be prepared to do the whole spectrum. That's what it's been trained to do anyway.

It was very interesting to see Hurricane Andrew and the response, how the people in Florida reacted gratefully to the military who were down there helping them. They had a lot of criticism about things locally, but they heaped praise on the military. These were military people who were not particularly trained in all cases for that sort of response.

The military must be ready for both traditional and non-traditional roles. To guarantee national security and foster a peaceful and prosperous world in which to conduct trade and commerce, these roles can be expected to play a large part in collective action. This does not abrogate U.S. leadership through collective action. In fact, it strengthens the bindings that the United States has with other nations.

Addendum

Comments from Senior Officials

"We may need less military,
but we won't need our military any less".

THE FINAL SESSION OF THE SYMPOSIUM WAS titled "Pros and Cons: Capabilities in Search of Missions." The membership of this panel was made up of senior officials representing organizations that play a major part in executing the non-traditional roles discussed throughout the symposium. First, Air Force Lieutenant General John B. Conaway, Chief of the National Guard Bureau, discusses the role of the National Guard in executing non-traditional missions. Next, Coast Guard Rear Admiral Robert E. Kramek, Chief of Staff, U.S. Coast Guard, provides the Coast Guard view. Navy Admiral Paul David Miller, Commander-in-Chief, United States Atlantic Command, shares his views on how the U.S. military needs to be reshaped to meet the changing international environment. Finally, Dr. Alberto Coll, Principal Deputy Assistant Secretary of Defense, Office of the Assistant Secretary of Defense for Special Operations and Low-Intensity Conflict, closed the presentations with a thought-provoking talk on non-traditional roles as instruments of policy formulation. Summaries of their comments follow. After the individual presentations, a short discussion period took place. Several sets of questions and answers reflecting the substance of the discussion period are provided at the end of this section.

THE ROLE OF THE NATIONAL GUARD

The many missions of the National Guard include support for the counterdrug effort, deploying the Guard in support of the Gulf War, disaster relief assistance, and potential Guard participation in such programs as the proposed Urban Youth Corps. The National Guard is a vital asset to any state and provides support to civil authorities in strict compliance with the applicable law. The missions of the National Guard have evolved during the years and now include many non-traditional roles.

THE COAST GUARD VIEW

The various Coast Guard functions are widespread and include many complexities of interservice/ agency operations. The Coast Guard is the only service capable of undertaking law enforcement functions. Additionally, the Coast Guard often finds it easier to understand non-traditional missions than do the other services in that the Coast Guard is not seen as a threat to the sovereignty of other nations. The Coast Guard's international reputation is based on its search and rescue operations. Indeed, the naval forces on many foreign nations look much like a small Coast Guard.

With the military drawdown of forces, the Coast Guard is increasing in numbers. It has new missions, and business is booming due to the non-traditional roles it has been assigned. The following recommendations were offered:

1) The Coast Guard roles need to be addressed in the JCS strategy; this is not currently the case;

2) Coast Guard roles and missions should be addressed in the Navy's War Plans and its new strategy, "From the Sea";

3) Emphasis must be placed on opportunities that allow the services to be mutually supportive, but non-redundant.

THE VIEW FROM THE ATLANTIC COMMAND

There is a need to fundamentally reshape the American military. The U.S. military force structure must be modified to meet the changing international environment. A smaller force will experience a decrease in strategic nuclear forces and an increased focus on regional conflicts, peacekeeping and support to domestic authorities. This force structure must be more capabilities-based than threat-based. As a smaller force, a premium must be placed on teamwork. In today's world, the military must function as both the sword and plowshare of U.S, national security policy. Moreover, in that non-traditional missions serve as excellent training for traditional military tasks, these missions will not serve to "dull the sword". The U.S. may well need less military, but it will not need the military any less.

SPECIAL OPERATIONS/LOW INTENSITY CONFLICT

Non-traditional roles are important and must be taken seriously as an instrument of policy formulation in the years ahead. The world situation is now more complex, but less manageable. Powerful regional actors are on the scene and will alter the face of international politics. The U.S. is not going to be the world's policeman, but it needs to act as a "senior managing partner" of the international system if it is to remain a global power with a degree of control over the international security environment. This will require United States diplomacy that is sensitive to regional differences. The U.S. (the military) will have to react to the actions of regional powers before U.S. national interests are actually threatened. Now, more than ever before, the U.S. must gain the greatest amount of political leverage from its military power. Non-traditional missions are thus part of the armory of missions the U.S. can use in shaping the world of today. These missions include: (1) The counter-proliferation of weapons of mass destruction and advanced weapon technologies; (2) Peacekeeping and peacemaking; (3) Foreign Internal Defense; and (4) Foreign disaster relief and humanitarian and civic action missions. There is also a need for the following:

(1) A rebalancing of missions and priorities. Do not assume that the capabilities required for non-traditional roles are any less than those required for traditional missions

(2) A discarding of the notion that non-traditional roles are simply chores that must be undertaken. These missions must be viewed as effective instruments of U.S. policy formulation

(3) Non-traditional missions must be addressed in U.S. national security policy

(4) Doctrine regarding non-traditional missions must be developed

(5) Greater interagency cooperation must occur

(6) Long-term planning for non-traditional missions must be accomplished

(7) The improvement of language and cultural skills of military personnel for regional defense purposes must become a priority.

In conclusion, the national security community must recognize the usefulness of non-traditional missions. The task of the military is to provide for the common defense. In today's world, all instruments of military power must be utilized, to include those of a non-traditional nature.

DISCUSSION FOLLOWING
THE PRESENTATIONS

QUESTION: If we are in the process of decreasing our military, it seems it would be in our best interest to decrease the size of other peoples' militaries. How does this square with our interest in foreign military sales? Also, why aren't we helping the Soviet Union retrain its military, or buying their weapons?

ANSWER: Counter-proliferation policies need rethinking. But, the reality of the situation is that all countries want modern weapons. In some countries, we are attempting to demonstrate the benefits of increased civilian control of the military. Also, we are emphasizing the benefits of having a home guard or a national guard in emerging countries, instead of a large offensive standing military.

QUESTION: In catastrophic national disasters, what variables must be considered before the governor asks for federalization of the troops?

ANSWER: The governor makes the decision to ask for federal assistance. In Florida, this was done three days after the hurricane hit. A review of the hurricane relief efforts in Florida reveals that FEMA responded about as fast as it possibly could. The great catastrophe in waiting to see what kind of damage assessment they could get, was that they couldn't get in by ground, and the air assessment didn't reveal all of the infrastructure losses. Perhaps the Governor should have requested Federal help when he flew over the area with the President, but, even from the air, the damage didn't seem as bad as it was.

The Guard was overwhelmed by the third day. They had the law enforcement mission. People started coming back to their homes. They were giving out thousands of MREs, providing humanitarian services, and got maxed-out. By the third day, the Guard was saying "help", and was telling the Governor that federal assistance was necessary.

QUESTION: Should our military force be threat-based or capabilities- based?

ANSWER: Our future force will be more capabilities-based than threat-oriented. We still need to deal with the threats on the horizon. But what is it that you want from the military forces? You want capabilities to be present, in forward positions, ready to be used by the Unified CINCs; not necessarily to counter threats, because they might occur in places where there are no threats, as we know "threats" in the historical sense. But you still want capabilities to employ, particularly in the mission areas this conference has been discussing.

When looking at "packaging" of forces, we need to send our CINC forces that are capability-centered, so that they can draw from that kit and use it for whatever mission comes up, to respond to any threats in those regions.

Reports from Discussion Group

DURING THE SYMPOSIUM, ALL PARTICIPANTS were assigned to seminar groups which met once each day for one hour each. The following pages summarize significant discussion points which were made during the discussion seminars. This part of the addendum is provided to help you to think about non-traditional roles and missions in more detail and to focus on what a sampling of the seminars considered to be the important issues.

Points of Discussion (Seminar A)

The following comments are derived from Professor Huntington's remarks:

- U.S. would not use nuclear weapons in response to chemical attacks or a single nuclear weapon against a third party.
- A preemptive strategy for dealing with a Saddam with nuclear weapons.
- Americans are defensive in nature—not offensive.

Comments based on remarks made by other speakers:

- The threat has fundamentally changed—not simply receded.
- Russia is still a latent military threat.
- Can we afford the cost of non-traditional roles plus the UN-Russia-Somalia?
- We need to build "Safeways" in Russia.
- We should help the Russians suppress internal conflicts in the former Soviet republics.

Points of Discussion (Seminar B)

This group examined primarily four issues:

- Definition of "non-traditional roles"
- What are non-traditional roles?
- Are non-traditional roles budget/force structure drivers/ploys?
- Are/What changes are required to deal with non-traditional issues?

Consensus of the group follows:

■ non-traditional roles are not new. We did them during the Cold War and before, and will continue to do them in the future. There was a sense that the end of the Cold War has shifted the focus—the same roles and missions are still necessary—the emphasis, and sometimes the scale, is different.

■ "New" emphasis on non-traditional roles should not cloud military's primarily role—to defend national interests on the field, on the seas, and in the air, of battle—when called upon.

■ Non-traditional roles are not budget/force structure drivers. Post-Cold War restructuring should address them as part of a "rational" plan for down-/right-sizing.

■ Military should emphasize non-traditional capabilities, as resources permit. This is particularly true in a period when warfighting readiness requirements are lessened from what we faced during the Cold War. The military has always been used in pursuit of national objectives, both in, and out, of combat.

Points of Discussion (Seminar C)

■ As U.S. military forces become more involved in post cold war disturbances in the international context, it is likely that they will have to deal with complex political, cultural, social, and religious issues which go well beyond the traditional combat roles.

■ The services are already doing a great deal of "non-traditional" training for new roles such as hostage rescue, drug enforcement, and non-combatant evacuation. The reserve component has a wide range of expertise in non-traditional roles and missions and this should be maintained and exercised regularly. However, frequent call up of these resources for less than national emergencies might result in reduced capability over time.

■ In spite of much rhetoric about the increasing importance of the UN in settling international disputes in recent years, the UN remains a very fragile and functionally weak organization (no standing forces, little funding, and no real authority to initiate peacemaking/keeping/enforcement operations on its own). If the U.S. truly believes the UN has a greater role to play in the international context of maintaining world order, then perhaps the U.S.ought to take the lead in re-examining its traditional policies which are perceived by many as supporting UN initiatives only when the U.S. interest is best served. If we want and expect the UN to play a major role, then we have to support the UN more fully to make it an effective and respected organization.

■ The most viable non-combatant role for military forces is disaster relief operations. Regardless of the extent of the drawdown in force structure, the residual U.S. military has on call the necessary skills, equipment, and supplies to respond quickly and effectively to natural disasters. However, this is not a new role as there are numerous examples where U.S. forces have been so engaged.

■ Other non-traditional roles and missions such as domestic nation re-building to include educating and training our underprivileged youth, people to people programs in our cities, and in other applications where the skills and capabilities can best be used were examined. The greatest limitation to effective employment of active duty military forces in these roles is the downsizing which is expected to continue at accelerated rates in the next few years. The remaining active force would be hard pressed to maintain its readiness for combat and provide large masses of people dedicated to nation re-building at the same time. It was noted that there is a tremendous pool of talent to meet these needs in the thousands of servicemen and women who will be leaving active duty. Many former military bases could be converted to support local needs for training and self development programs under the aegis of local, state, or non-DoD federal sponsorship.

Points of Discussion (Seminar D)

The following questions were asked and then discussed:

■ Should the primary role of the military services, as defined by Title 10 of the U.S. Code, be amended to include non-traditional roles such as peacekeeping, peacemaking, humanitarian intervention, disaster relief and domestic assistance during civil unrest?

■ Should members of the services be trained, equipped and sized for these roles?

■ The consensus of the group was that the U.S. armed forces should participate in these activities as long as they do not interfere with the services' primary responsibilities of combat and combat readiness. The group felt that these activities were both useful to society and serve as excellent training opportunities for members of the military services. These activities, however, should be assigned to the services as missions, not as Title 10 roles. The group felt that Title 10 already allows for these missions under the clause: "and other duties assigned by the president". They also felt that if these activities were specifically mandated by law, the services would lose the flexibility they now have with respect to training, equipping, and structuring their forces.

■ A concern was raised that since the U.S. military is the most powerful in the world, it serves as a model to others. This implies that any action the U.S. military takes may set a precedent and be used to justify the actions of other militaries. If the U.S. armed services become more involved in domestic affairs, no matter how noble the reason, this action may be interpreted by others that it is acceptable to use the military as a tool of domestic authority and legitimacy. While the U.S. may justify its actions based on the doctrine of strict civilian control of its military and its benevolent intentions, others may not meet these conditions thereby perpetuating the concept that a military-run government is acceptable.

■ Citing the differences between domestic and international roles for the military, the group focused its discussion on domestic roles. For these activities they felt that the lead should be given to the states' National Guard units. This arrangement is desirable for numerous reasons. First, it helps lessen the concern just raised by using state rather then federally controlled troops. Secondly, many of the non-traditional roles of the military are, in fact, traditional roles of the National Guard. Therefore, they can be trained, equipped and structured specifically for these roles. Also, since they are not subject to the restrictions imposed by Posse Comitatus they can act as law enforcement officers when necessary. Finally, if the National Guard is prepared for these roles, when the situations arise that require the support of the active military, the military commanders can use the Guard troops as their advisors and liaisons with the civilian authorities. This reduces the regular military's need to be trained and equipped for these roles, while preserving their ability act quickly, decisively and legally when called upon.

■ In summery, the group felt that the U.S. military forces should participate in non-traditional activities, but not at the expense of their traditional roles. Therefore, doctrine, leadership and training should be developed and forces should be equipped for these activities. Extreme care should be taken, however, to ensure that the responsibilities for these activities are placed with the most appropriate organizations.

Points of Discussion (Seminar E)

The basic conclusions of the group discussion follow:

- Non-traditional activities in foreign lands must meet specific U.S. national interests in order to gain public support.
- The UN may often serve as a convenient multilateral "cover" for activities which are in U.S. national interest but might not be welcome if done as unilateral U.S. activities.
- Conference has not addressed technology as a non-traditional response. For example, use electronic stream to knock out all public radios in Serbia.
- Oftentimes—as in the past—non-traditional problems are not best solved by military forces. Interagency coordination and the use of the proper national resource (aid, agriculture) should be considered and used.

About the Editor

James R. Graham is the Director of Symposia in the Institute for National Strategic Studies of the National Defense University. Before his retirement in 1992 from the U.S. Air Force, Colonel Graham served as the Senior Air Force Fellow and Deputy Director of the Strategic Capabilities Assessment Center at Fort McNair. Highlights of his Air Force career follow:

Colonel Graham completed Officer Training School and was commissioned as a second lieutenant in 1967. In 1968, he was assigned as Chief of the Ground Communications Operations Branch for Headquarters, U.S. Air Force Security Service in San Antonio, Texas. In 1969, he traveled to West Berlin, Germany, where he was the Communications Officer in the 6912th Security Squadron. In 1973, he returned to the states as a Section Commander at the Air University's Squadron Officer School. After attending the Air Force Telecommunications Staff Officer course in 1977 and 1978, Colonel Graham was assigned to Offutt Air Force Base, Nebraska, where he served as Director of Technical Plans, Strategic Communications Division, Air Force Communications Command. In 1981, he commanded the Systems Engineering and Support Squadron, a NATO organization providing telecommunications and computer support to the 6ATAF and LANDSOUTHEAST in Izmir, Turkey. He returned to Nebraska in 1982 to serve as Chief, Strategic Programs Development Division at Strategic Air Command. Colonel Graham was then assigned to the Pentagon as Director for Information and Resources Management for the Joint Staff.

Colonel Graham received a B.A. from Wabash College, an M.S. from Wayne State University, and an M.A. from the University of Southern Mississippi. In 1981 he graduated from the Air War College. He is also a graduate of the Air Command and Staff College, the National Security Management Course, and Squadron Officer School.

133

★U.S GOVERNMENT PRINTING OFFICE 1993-355-783/90031